Clyde Arthur Morrison

Crystal Fields
for Transition-Metal Ions
in Laser Host Materials

Springer-Verlag

Berlin Heidelberg New York
London Paris Tokyo
HongKong Barcelona Budapest

Dr. Clyde Arthur Morrison
Harry Diamond Laboratories
2800 Powder Mill Road
Adelphi, MD 20783 - 1197
USA

ISBN 3-540- 55465-3   Springer-Verlag Berlin Heidelberg New York
ISBN 0-387- 55465-3   Springer-Verlag New York Berlin Heidelberg

Typesetting: Camera ready by author
51/3020-5 4 3 2 1 0 - Printed on acid -free paper

Foreword

The original impetus for a methodical analysis of the spectra of transition-metal ions in crystalline solids originated with the late Nick Karayianis, my long-time coworker, whom I sorely miss. This work would never have been completed without the help of Donald Wortman, who cleared the administrative paths, oversaw the acquisition of various word processors, and gave excellent technical advice, all of which I very much appreciate. I wish to thank Richard Leavitt, John Bruno, and Gregory Turner for the development of the crucial computer programs and for simplifying the operation of these programs.

During my first year in graduate school, my sister Helen and her husband the late W. Edward Sanborn, came to my financial aid, without which I would not have been able to finish. Thank you, Helen.

Finally, I would like to thank my immediate family, Maria and Ken, Scott and Tina, and Tom and Amy for their warmth and encouragement during my recent illness. Most of all during this period my wife, Sue, was her splendid self always there with the right thing at the right time—care, love, and encouragement.

Contents

Tables

1.   Introduction

This book presents the experimentally determined parameters of the $3d^N$, $4d^N$, and $5d^N$ transition-metal ions in the second, third, and fourth ionization states in a number of different hosts. Although the parameters obtained by fitting the experimental optical levels in these hosts were originally reported in a variety of different notations, in this book these parameters have been converted to Slater parameters and crystal-field parameters that describe the crystal-field interaction in a spherical tensor form.

The host materials presented were predominantly chosen from past and present compounds doped with transition-metal ions that were shown to lase or show promise of being good laser materials, such as Al_2O_3. Many of the compounds chosen were investigated as possible host materials a number of years ago, but were discarded before lasers began to be used as pumps. A recent example of a laser-pumped laser is the titanium-doped Al_2O_3 laser that is so popular at present. It would not be surprising if some of these older compounds could serve as hosts for transition-metal ions for laser-pumped tunable lasers in the future.

It is impossible to cover all the host materials that have been reported in the literature. The field is narrowed somewhat by avoiding materials in which the transition-metal ion has been assumed to replace a constituent ion of a different ionization state, and no attempt was made to compensate for the charge imbalance during the crystal growth. Most of the materials included have had optical absorption and emission data reported. The pertinent crystallographic data and the detailed x-ray data on each host are included in the notation of the *International Tables* (Henry and Lonsdale, 1969). Thus the point group symmetry of the site assumed to be occupied by the transition-metal ion is known, and the reported data are analyzed with the site symmetry taken into account. For most of the host materials, a query by a fellow research worker triggered the investigation of a particular compound. In order to answer this query, a more or less complete bibliography of literature on the particular compound was accumulated, and a number of computations were performed in order to prepare an answer.

In the tables that make up the bulk of this work, the reported "free-ion" parameters (B, C, F_k, etc) are all converted to the Slater parameters, $F^{(k)}$, primarily because the Hartree-Fock calculations and the free-ion experimental data are predominantly reported in terms of $F^{(k)}$. Also, in the fitting of experimental data, it is easier to maintain the restriction $F^{(2)} > F^{(4)}$. Formulas are given for the conversion of the numerous parameters reported in the literature to the Slater parameters.

For the analysis of the spectra of the nd^N ions in a solid, the choice of Dq for the strength of the cubic crystal fields is an excellent one. However, when the ion occupies a site of lower symmetry, the number of different notations becomes excessive. Even for the sites with C_3 or C_4, the number of different notations is large, and the phases (+/–) are ambiguous. This causes the conversion from one set to another by the use of textbook formulas to be extremely hazardous. After considerable effort to try to settle on one type, we decided that the easiest way to express the results of fitting the experimental data was the spherical tensor method with the associated crystal-field parameters B_{nm}. This decision was not made without prejudice; it was affected by our background experience in the analysis of the rare-earth ion spectra where the B_{nm} are used almost exclusively. In the conversion of the various reported parameters, "standard" conversion formulas were used and the energy levels compared with theoretical energy levels if reported. If the energy levels did not compare with the corresponding reported values, various phases were tried in the conversion formulas; if this failed, either the reported theoretical or experimental values were least squares fit by varying the B_{nm} to obtain a best fit. If the theoretical energy levels are given, this procedure almost always gave a fit to within differences introduced by roundoff.

In the tables, a number of compounds are included for which only detailed x-ray data are found. These have been included because the site symmetry is of interest, and estimates of the crystal-field strength based on the point ion model show promise of a potential new laser. The tables do not include other important factors, such as crystal hardness, optical quality, and thermal conductivity; but available data on these quantities are given in the references. References to the index of refraction and electron spin resonance results are also included.

2. Theory and Data

2.1 Free-Ion Parameters

In the analysis of the experimental data on the transition-metal ions with the d^N configuration, the following Hamiltonian is used:

$$H_{F-I} = \sum_{k=2,4} F^{(k)} h_k + \alpha L(L+1) + \zeta \sum_{i=1}^{N} \mathbf{l}_i \cdot \mathbf{s}_i \qquad (1)$$

with

$$h_k = \sum_{i,j>i}^{N} \sum_{q=-k}^{k} C_{kq}^{*}(i) C_{kq}(j)$$

and

$$C_{kq}(i) = \sqrt{\frac{4\pi}{2k+1}} Y_{kq}(\theta_i, \phi_i) \ .$$

Also

$$C_{kq}^{*} = (-1)^{q} C_{k-q} \ .$$

The parameters $F^{(k)}$, α, and ζ are obtained by fitting the experimental data. The term involving α was introduced by Trees (1964) in his analysis of the spectrum of praseodymium. Frequently additional interactions are included with corresponding parameters that improve the fit to the experimental data. Since the purpose here is to present the data on the transition-metal ions in a solid, equation (1) applies to both free ions and ions in a solid. Data on the free ions with the $3d^N$ electronic configuration in an extensive number of ionization states have been reported by workers at the National Institute of Science and Technology (NIST, formerly the National Bureau of Standards) in the Journal of Physical and Chemical Reference Data.

The parameters of equation (1) for the free ions of the $3d^N$ transition-metal ions in the doubly, triply, and quadruply ionized state are given in table 1, part A. In general, the parameters from different sources vary only slightly, but the most consistent set is that of Uylings et al (1984), which is an analysis of the most recent experimental data and presents the entire $3d^N$ configuration for $N = 2$ to 8. The set of parameters reported by Pasternak (1972) for the doubly ionized $3d^N$ configuration is convenient for computations involving the entire series, in that algebraic expressions for $F^{(k)}$, α, and ζ are given as functions of N.

The experimental data on the $4d^N$ series are less abundant; for these ions, the parameters of equation (1) are given in table 1, part B. For this series the reported parameters vary considerably; in particular, those reported by Di-Sipio et al (1970) differ from the more complete analysis of other workers for the $4d^8$ configuration, but theirs are the only readily available values for the $4d^5$ configuration.

For the transition-metal ions with the $5d^N$ electronic configuration, the number of experimental values of $F^{(k)}$, ζ, and α is very small (only 10). The parameter values obtained by fitting the experimental data are given in table 1, part C. The outstanding characteristic of this series is the tremendous increase in the spin-orbit constant, ζ, in going from the $3d^N$ series to the $5d^N$ series. For example, for the triply ionized nd^2 configuration, the value of ζ increases by greater than a factor of 10 for Ta^{3+} $(5d^2)$ over V^{3+} $(3d^2)$. On the other hand, the values of $F^{(2)}$ and $F^{(4)}$ decrease in going from V^{3+} to Ta^{3+}. Thus, while the spin-orbit interaction can be ignored in the $3d^N$ configuration, it must be taken into account in the $5d^N$ series. The parameters of the $4d^N$ series are intermediate to these two configurations; depending upon the particular case being investigated, a qualitative analysis of the experimental data will favor which choice to use.

Hartree-Fock values for three ionization states (doubly, triply, and quadruply ionized) of the $3d^N$, $4d^N$, and $5d^N$ configurations are available (Fraga et al, 1976), and these are given in table 2. Also in table 2, the Hartree-Fock values for $<r^2>$ and $<r^4>$ are given. A comparison of the $F^{(k)}$ in table 1 with the corresponding Hartree-Fock values of table 2 shows that $F^{(k)} \approx 0.8\, F^{(k)}_{HF}$ and $z \approx 0.9\, z_{HF}$. Since the $F^{(k)}$ vary as $<1/r>$, this would indicate that the Hartree-Fock wavefunctions are too large near the nucleus. Consequently, the Hartree-Fock values of $<r^k>$ are expected to be too small. An approximate correction for this defect (which also occurs in the rare-earth ions) has been to introduce an expansion parameter, τ, such that $<r^k> = <r^k>_{HF}/\tau^k$ (Morrison, 1988, p 125).

The free-ion interaction given in equation (1) involving the Slater integrals is frequently expressed in terms of different parameters, and the parameters obtained by fitting the energy levels of the ions in a solid are expressed in terms of these parameters. Two sets of parameters are the Slater parameters, F_k, and the Racah parameters, E^k. Because the matrix elements of the Coulombic interaction are given in Nielson and Koster (1963) as multipliers of $F^{(k)}$, we have chosen to put all the reported data in terms of $F^{(k)}$. Judd (1963) gives the following relations:

$$F^{(2)} = 49F_2 \ ,$$
$$F^{(4)} = 441F_4 \ , \tag{2}$$

and

$$F^{(2)} = 49(9E^2 - E^1)/2 \ ,$$
$$F^{(4)} = 441(5E^2 - E^1)/10 \ . \tag{3}$$

In terms of B and C, introduced by Racah (1949), the $F^{(k)}$ are

$$F^{(2)} = 7(7B + C) \ ,$$
$$F^{(4)} = 63C/5 \ . \tag{4}$$

These three relations (eq (2), (3), and (4)) are the most common parameters reported from the analysis of experimental data. By far the most frequent parameters are B and C, where much of the experimental analysis has been facilitated by the use of the energy diagrams of Tanabe and Sugano (1954). The spin-orbit interaction in the lowest energy LS term (Hund ground state) is frequently represented by

$$H_{s-o} = \lambda \mathbf{L} \cdot \mathbf{S}, \tag{5}$$

where $\mathbf{L}$ is the total angular momentum operator and $\mathbf{S}$ is the total spin angular momentum operator. The relation to ζ is

$$\zeta = \pm 2S\lambda, \tag{6}$$

where the sign of λ is positive for $1 \leq N \leq 4$ and negative for $6 \leq N \leq 9$ for the d^N configuration ($\zeta > 0$ for all N).

It has been known for some time that the parameters $F^{(k)}$ are decreased when the ion enters a solid (Figgis, 1966, ch. 9). The most recent expression for this reduction is that of Morrison (1980):

$$\Delta F^{(k)} = -\left\langle r^k \right\rangle^2 S^{(k)}, \tag{7}$$

where

$$S^{(k)} = -e^2 \sum_i \frac{\alpha_i Z_i (k + 1)}{R_i^{2k+4}} \tag{8}$$

and Z_i is the number of ligands with polarizability α_i located at R_i. An expression for the interconfiguration shift $nl^N - nl^{N-1} n'l'$ is also given as

$$\Delta E_{nln'l'} = \Delta E_o - \left[\left\langle r^2 \right\rangle_{nl'} - \left\langle r^2 \right\rangle_{nl} \right] S^{(0)} \tag{9}$$

with $S^{(0)}$ given in equation (8). A shift in the spin-orbit constant was also given as

$$\Delta \zeta = -\frac{(\alpha_o a_o)^2 e^2}{2} \sum_{n=1}^{\infty} n(n+1) \left\langle r^{2n-2} \right\rangle \sum_i \frac{Z_i \alpha_i}{R_i^{2n+4}} \tag{10}$$

where α_o (~1/137) is the fine structure constant and a_o (~0.5 Å) is the radius of the Bohr orbit. The shift in the spin-orbit constant given in equation (10) is negligible even for ligands where R_i is small. For six ligands at $R = 1.6$ Å and $\alpha \approx 1$ Å^3, $\Delta\zeta \approx -0.5$ cm^{-1}, which can be ignored.

In fitting the spectra for d^N ions in a solid, the generally accepted ranges of the parameters are $F^{(2)} > F^{(4)}$, $0 < \alpha < 100$ cm^{-1}, and ζ less than or equal to the free-ion value and positive. The condition for $F^{(2)} > F^{(4)}$ can be proven from the definition of the Slater integrals, and the condition for $\alpha > 0$ appears to come from values obtained by fitting the free-ion spectra. The condition that $\zeta > 0$ gives considerable leeway in the fitting; however, several papers have reported $\zeta < 0$ (not λ in eq (6)), which is nonsense considering the definition of ζ. If the result given in equation (10) is the only mechanism considered to affect the spin-orbit interactions, then one would assume that the values of ζ should be near the free-ion value.

2.2 Crystal-Field Parameters

In the phenomenological theory of crystal fields, the interaction of the electrons in the d^N configuration with the crystal field is taken as

$$H_{CEF} = \sum_{nm} B_{nm}^* \sum_{i=1}^{N} C_{nm}(i), \qquad (11)$$

where $n = 2$ and 4, $-n \le m \le n$, and the crystal-field parameters obey the relation $B_{nm}^* = (-1)^m B_{n,-m}$. The $C_{nm}(i)$ are related to the spherical harmonics by equation (1). The phenomenological method consists in finding those B_{nm} that best fit the experimental data for the ion under investigation, much the same as the Slater parameters are obtained by fitting the free-ion data. The number of crystal-field parameters in equation (11) is determined by the symmetry of the site, and Morrison (1988, ch. 8) gives the method of finding the nonvanishing B_{nm} for the 32 point groups. For the groups C_2 through O_h, table 1d gives the number of crystal-field parameters, B_{nm}, or the number of crystal-field components, A_{nm}, for each value of n. For sites with C_1 or C_i symmetry, there are five crystal-field parameters for B_{2m} (which can be reduced to four by an appropriate rotation) and nine crystal-field parameters B_{4m}. The determination of such a large number of crystal-field parameters requires the acquisition of a large number of experimental levels. Thus, if we include only the Slater parameters $F^{(2)}$ and $F^{(4)}$ in the fitting, we have 15 parameters, and we would need more than 15 experimental levels to attempt a sensible fit. This requirement practically excludes consideration of a fitting of data taken on ions occupying sites with C_1 or C_i symmetry. Even in C_2 symmetry, there are seven crystal-field parameters, which generally exceeds the number of experimental levels. For sites with C_{2v} symmetry, the number of crystal-field parameters is five, which is excessive, and for C_3 or C_4 symmetry there are only three B_{nm} (B_{20}, B_{40}, and B_{43} for C_3; and B_{20}, B_{40}, and B_{44} for C_4). Obviously, for the low symmetry sites (lower than C_3 or C_4), some approximation must be made before one can start any methodical fitting of the experimental data.

The most common assumption made in the experimental data is that the site has approximately cubic symmetry. The experimental data are then analyzed by using the appropriate Tanabe-Sugano (1954) energy diagram. Justifying the cubic approximation requires considerable experience. Figgis (1966, ch. 9) has a detailed discussion of the analysis of experimental spectra by the use of Tanabe-Sugano diagrams. Also, Leihr (1963) gives Tanabe-Sugano plots for $3d^3$ and $3d^7$, including the spin-orbit interaction. To use the Tanabe-Sugano diagrams, we need the relation of B_{nm} of equation (11) with the Dq used in these diagrams.

2.3 Relation of Dq with B_{40}

McClure (1959) gives the potential energy of a single electron in a sixfold cubic array of charges at a distance R as

$$U = \frac{105}{2} Dq \left(X^4 + Y^4 + Z^4 - \frac{3}{5} \right) . \tag{12}$$

where $X = x/r$, $Y = y/r$, and $Z = z/r$.

In the notation of equation (11) (for sixfold cubic coordination with charges at $(\pm R,0,0)$, $(0,\pm R,0)$, and $(0,0,\pm R)$), we write the same potential energy as in equation (12) as

$$U = B_{40} \left[C_{40} + \sqrt{\frac{5}{14}} \left(C_{44} + C_{4-4} \right) \right] . \tag{13}$$

The C_{nm} are given by (Morrison, 1988, p 14)

$$C_{40} = \left(35Z^4 - 30Z^2 + 3 \right)/8 , \tag{14}$$

$$C_{4\pm4} = \left(X \pm iY \right)^4 \left(35/128 \right)^{1/2} . \tag{15}$$

Substituting equations (14) and (15) into equation (13) gives

$$C_{40} + \sqrt{\frac{5}{14}} (C_{44} + C_{4-4}) = \frac{5}{2} \left(X^4 + Y^4 + Z^4 - \frac{3}{5} \right) . \tag{16}$$

Thus, we obtain

$$\frac{5}{2} B_{40} = \frac{105}{2} Dq ,$$

or

$$B_{40} = 21\,Dq \ ,$$
$$B_{44} = \sqrt{5/14}\;B_{40} \ . \tag{17}$$

This result, equation (17), has been used to convert the Dq reported in the literature to B_{40}. For the above type of cubic symmetry and for the d^N configuration, we write equation (11) as

$$H_{CEF} = B_{40} \sum_{i=1}^{N} \left[C_{40}(i) + \sqrt{\frac{5}{14}} \left(C_{44}(i) + C_{4-4}(i) \right) \right] \ . \tag{18}$$

If, in the cubic group, the principal axis of rotation is the cube diagonal, then the crystal-field interaction is

$$H_{CEF} = B_{40} \sum_{i=1}^{N} \left[C_{40}(i) + \sqrt{\frac{10}{7}} \left(C_{43}(i) - C_{4-3}(i) \right) \right] \tag{19}$$

with $B_{40} = 14\,Dq$, and

$$B_{43} = \sqrt{10/7}\;B_{40} \ .$$

If we write the cubic field parameter in equation (19) as $B_{40}^{(3)}$ and the cubic field parameter in equation (18) as $B_{40}^{(4)}$, then for the *same* crystal field in the two descriptions we have

$$B_{40}^{(3)} = -\frac{2}{3}\,B_{40}^{(4)} \ . \tag{20}$$

Thus, in octahedral symmetry we have $B_{40}^{(4)} > 0$ and $B_{40}^{(3)} < 0$.

A second method is to calculate the crystal-field components, A_{nm}, obtained by using the detailed x-ray data of the crystal under investigation. The monopole crystal-field components, $A_{nm}^{(0)}$, are given by

$$A_{nm}^{(0)} = -e^2 \sum_{j} \frac{q_j C_{nm}(j)}{R_j^{n+1}} \tag{21}$$

(Morrison and Leavitt, 1982; Karayianis and Morrison, 1973) where q_j is the charge on the ligands (in units of electron charge) at R_j and the sum covers all the ions in the solid. The consistent dipole contribution to the crystal-field components, $A_{nm}^{(1)}$, is given by

$$A_{nm}^{(1)} = -e^2 \sum_{\mu} \sqrt{(n+1)(2n+3)} \left\langle 1(\mu) n + 1 (m - \mu) \middle| n(m) \right\rangle$$
$$\times \sum_{j} \frac{Q_{1\mu}(j) C_{n+1, m-\mu}(j)}{R_j^{n+2}} \tag{22}$$

(Morrison, 1976, and the work of Faucher and Garcia, 1982, 1983; Garcia and Faucher, 1984; and Garcia, Faucher, and Malta (1983)) where $Q_{1m}(j)$ is the electric dipole moment of the ion at R_j, which is determined self-consistently. The quantity in angular brackets is a Clebsch-Gordan coefficient (Morrison, 1988). A third contribution to the crystal-field components is the self-induced effect given by

$$A_{nm}^{S-I} = -e^2 \sum_{j} a_j \frac{(n+1)!}{n!} \left[2^{2n-1} \frac{(n!)^2}{(2n)!} - 1 \right] C_{nm}(\widehat{R}_j) / R_j^{n+4} \quad , \tag{23}$$

(Morrison et al, 1982) where α_j is the polarizability of the ions at R_j. The total crystal-field components are $A_{nm} = A_{nm}^{(0)} + A_{nm}^{(1)} + A_{nm}^{S-I}$. In the point ion crystal-field theory, the crystal-field parameters B_{nm} are given by

$$B_{nm} = \rho_n A_{nm} \quad , \tag{24}$$

where ρ_n represents the effective value of $<r^n>$. In the analysis of the spectra of rare-earth ions, the ρ_n have been found to be dependent on the rare-earth ion and independent of the particular host (Morrison and Leavitt, 1979), a situation which does not seem to be true for the transition-metal ions. In general, using Hartree-Fock values for $<r^n>$ in equation (24) gives values of B_{nm} too small. Nevertheless, when the reported B_{nm} obtained by fitting the experimental data are compared with the monopole A_{nm}, the signs agree in practically all cases.

To find the effective ρ_n values in equation (24) to be used in cases where the site symmetry is lower than cubic, we assume first that the cubic approximate B_{40} has been obtained by using the appropriate Tanabe-Sugano diagram. The rotational invariants $S_n(A)$ and $S_n(B)$ (Leavitt, 1982) are calculated by using

$$S_n(X) = \left[X_{n0}^2 + 2 \sum_{m>0}^{n} X_{nm}^* X_{nm} \right]^{1/2} \quad , \tag{25}$$

where X_{nm} is either B_{nm} or A_{nm}.

Then by equation (24) we have

$$\rho_n = S_n(B)/S_n(A). \tag{26}$$

For the cubic field described along the cubic (001) axis or along the (111) axis (or any other axis),

$$S_4(B) = 6\sqrt{21}\, Dq \; . \qquad (27)$$

with Dq determined for the experimental data. Thus we see that within a constant factor, Dq is a rotational invariant—a fact that contributes to its success in characterizing the strength of the crystal field.

The $S_4(A)$ can be calculated from the crystal-field components for the correct symmetry by using equation (25). An approximate set of crystal-field parameters, B_{nm}, for the ion at the particular site can then be calculated using the A_{nm} for the correct symmetry and the ρ_n from equation (26). The resulting B_{nm} can be used to calculate the energy levels; these levels can then be compared with experimental values. This procedure avoids the extremely difficult method of viewing the crystals from numerous different angles and attempting to determine the similarity to cubic symmetry. This process of finding improved calculations of the energy levels when the site occupied by the transition-metal ion has lower symmetry has been applied to the analysis of Cr^{3+} in the C_{3i} site of $Gd_3Sc_2Ga_3O_{12}$ (Gruber et al, 1988), of Fe^{3+} in the garnets $R_3Al_5O_{12}$ (R = Gd, Tb, ...,Yb, Lu) (Morrison et al, 1987), and of the spectra of triply ionized $3d^N$ ions in both the C_{3i} (octahedral) site and the S_4 (tetrahedral) site in $Y_3Al_5O_{12}$ (Morrison and Turner, 1987). In all cases, this method gave an improved representation of the experimental results.

2.4   Relation of B_{nm} to Other Notations

For a crystal field of low symmetry, the correlation of the various notations used in the analysis of the crystal-field interaction is extremely difficult, and we shall not attempt to be complete here. Instead we shall relate what appears to be the most prevalent. For a crystal field of C_4 and higher symmetry, we write the crystal-field interaction as

$$H_{CEF} = B_{20} \sum_{i=1}^{N} C_{20}(i) + B_{40} \sum_{i=1}^{N} C_{40}(i) + B_{44} \sum_{i=1}^{N} \left[C_{44}(i) + C_{4-4}(i) \right] \; , \quad (28)$$

where

$$C_{nm}(i) = \sqrt{4p/(2n+1)}\; Y_{nm}(q_i, f_i)$$

and all B_{kq} can be taken real. The choice of B_{kq} real is convenient in fitting, but in comparing these parameters with theoretical B_{kq}, we may have to rotate the calculated parameters about the principal axis (see Morrison, 1988, for a detailed discussion).

The matrix elements of equation (28) for a single electron are given by

$$<20|H_{CEF}|20> = \frac{2}{7} B_{20} + \frac{2}{7} B_{40} \ ,$$

$$<2\pm1|H_{CEF}|2\pm1> = \frac{1}{7} B_{20} - \frac{4}{21} B_{40} \ ,$$

$$<2\pm2|H_{CEF}|2\pm2> = \frac{-2}{7} B_{20} + \frac{1}{21} B_{40} \ , \qquad (29)$$

$$<2\pm2|H_{CEF}|2\mp2> = \frac{\sqrt{70}}{21} B_{44} \ ,$$

(Morrison and Leavitt, 1984), where $|2m> = Y_{2m}$, $<2m'|C_{kq}|2m> = \int Y^*_{2m'} C_{kq} Y_{2m}$; $d\Omega$ with $d\Omega = \sin \theta d\theta d\phi$, and all the arguments of Y_{2m} and C_{kq} are θ and ϕ.

Ballhausen (1962, p 101) gives the corresponding matrix element for tetragonal fields (eq (28)), in terms of Dq, Ds, and Dt. Thus, the following relations exist:

$$B_{20} = -7Ds \ ,$$

$$B_{40} = 21(Dq - Dt) \ , \qquad (30)$$

$$B_{44} = 3\sqrt{70} \, Dq/2 \ .$$

By comparing the matrix elements of Griffith (1961) for tetragonal symmetry, we obtain

$$B_{20} = \delta - \mu \ ,$$

$$B_{40} = 21Dq - (\delta + 3\mu/4) \ , \qquad (31)$$

$$B_{44} = \sqrt{70} \left[3Dq + \frac{\delta + 3\mu/4}{5} \right]/2 \ .$$

For C_3 (trigonal symmetry) and higher symmetry, we write the crystal field for the electronic configuration d^N as

$$H_{CEF} = B_{20} \sum_{i=1}^{N} C_{20}(i) + B_{40} \sum_{i=1}^{N} C_{40}(i) + B_{43} \sum_{i=1}^{N} [C_{43}(i) - C_{4-3}(i)] \qquad (32)$$

The matrix elements of the first two terms of equation (32) are the same as in equation (29) above, and

$$\langle 2-2|H_{CEF}|21\rangle = \sqrt{35} B_{43}/21 \ . \qquad (33)$$

By comparing the matrix elements for trigonal symmetry of Ballhausen (1962, p 104), we obtain

$$B_{20} = -7D\sigma \ ,$$
$$B_{40} = -14Dq - 21D\tau \ , \qquad (34)$$
$$B_{43} = 2\sqrt{70} \ Dq \ .$$

Similarly for the parameters of Pryce and Runciman (1958), we obtain

$$B_{20} = v - 2\sqrt{2} \ v' \ ,$$
$$B_{40} = -14Dq + 2w/3 \ , \qquad (35)$$
$$B_{43} = -\sqrt{7/10} \ (20Dq + w/3) \ ,$$

where

$$w = 2v + 3\sqrt{2} \ v'.$$

In obtaining the result given in equation (35), we use the equations given by Macfarlane (1963). In Macfarlane's convention, the B_{43} is negative; this has no effect on the energy levels, but introduces a phase factor in the wave function. In the tables we frequently change the sign of B_{43} to conform with the sign of the monopole A_{43} (B_{43} is obtained from equation (35) using values of Dq, v, and v' reported in the literature).

In all the relations given above (eq (30), (31), (34), and (35)), certain phase conventions are assumed; these equations *cannot* be used blindly. In order to obtain consistent values for the B_{nm}, we have frequently had to resort to changing the signs of some of the reported parameters. In some cases where theoretical levels are reported, we have resorted to fitting the theoretical results to obtain B_{nm}. Frequently, the point-charge crystal-field components, A_{nm}, indicate the correct phase relations and are used to determine the phases reported in the tables. A number of other notations for crystal-field parameters have been given by König and Kramer (1977).

2.5   Presentation of Data

Each host is described in a series of tables organized as follows.

2.5.1   Crystallographic data

The crystallographic data on each host are given in the notation of the *International Tables* (Henry and Lonsdale, 1969). The crystallographic data are presented in a short table for each host that lists the following information:

(a)   The crystal class: triclinic, orthorhombic, etc.

(b)   The space group symbol and number from the *International Tables*.

(c)   The number of chemical formula units, Z, per unit cell.

(d)   The setting, if there is more than one for that space group in the *International Tables*.

(e)   The position (site type in the *International Tables*), site symmetry (in the Schoenflies notation), and general x, y, and z coordinates (expressed as fractions of the lattice constants) for that site type, for each constituent of the host crystal.

(f)   The lattice constants a, b, and c (in angstroms) and angles α, β, and γ (in degrees and decimal parts).

(g)   The effective charges (usually the valence charge) in units of the electronic charge.

(h)   The electric-dipole polarizabilities, α (in Å^3), for each of the constituent ions.

2.5.2   Crystal-field components, A_{nm}, and parameters, B_{nm}

For each host, the data described in section 2.1 were used to obtain the point-charge equation (21), point-dipole equation (22), and self-induced equation (23) contributions to the crystal-field components, A_{nm}. All the A_{nm} for $1 \leq n \leq 5$ are given and are sufficient for the analysis of the nd^N configuration. The units of A_{nm} are $\text{cm}^{-1}/\text{Å}^n$. In a number of tables the crystal-field parameters for a particular ion are given by $B_{nm} = <r^n>A_{nm}$, where $<r^n>$ is the Hartree-Fock radial expectation value (Fraga et al, 1976) of r^n for the ion under consideration. At the bottom of a number of the tables of A_{nm}, the sums $S^{(0)}$, $S^{(2)}$, and $S^{(4)}$ from equation (8) are given.

2.5.3   Experimental results

Each host includes tables reporting all the experimental data in terms of the Slater integrals, $F^{(k)}$, and the crystal-field parameters, B_{nm}. The B_{nm} are calculated using equations (30), (31), (34), or (35). If the parameters are different from these, the reported theoretical or experimental levels are fitted using equations (28) or (32) to obtain the crystal-field parameters B_{nm}.

2.5.4   Bibliographies and reference lists

The final section on each host material consists of a bibliography of experimental and theoretical work that has been reported. These lists, in most cases, are far from exhaustive and should be augmented by a search of the recent literature. A number of hosts have only x-ray data reported, and no reference to optical data on transition elements in these hosts has been found. On a number of host materials, references were found that contain important information on that host not contained in the tables. These references have been included.

14

2.6 References

C. J. Ballhausen, *Ligand Field Theory*, McGraw-Hill, New York (1962), p 93.

M. V. Eremin and A. A. Kornienko, *Effect of Covalency on Slater Parameters and the Correlation Crystal Field in Transient-Metal Compounds*, Opt. Spectrosc. **53** (1982), 45.

M. Faucher and D. Garcia, *Electrostatic Crystal-Field Contributions in Rare-Earth Compounds with Consistent Multipolar Effects: I.—Contribution to K-Even Parameters*, Phys. Rev. **B26** (1982), 5451.

M. Faucher and D. Garcia, *Crystal-Field Effects on 4f Electrons: Theories and Realities*, J. Less-Common Metals **93** (1983), 31.

B. N. Figgis, *Introduction to Ligand Fields*, Interscience, New York (1966).

S. Fraga, K.M.S. Saxena, and J. Karwowski, *Physical Science Data: 5. Handbook of Atomic Data*, Elsevier, New York (1976). (See also Morrison and Schmalbach, 1985.)

D. Garcia and M. Faucher, *Crystal-Field Parameters in Rare-Earth Compounds: Extended Charge Contributions*, Phys. Rev. **A30** (1984), 1730.

D. Garcia, M. Faucher, and O. Malta, *Electrostatic Crystal-Field Contributions in Rare-Earth Compounds with Consistent Multipolar Effects: II.— Contribution to K-Odd Parameters (Transition Probabilities)*, Phys. Rev. **B27** (1983), 7386.

J. S. Griffith, *The Theory of Transition-Metal Ions*, Cambridge University Press, Cambridge, England (1961), p 226.

J. B. Gruber, M. E. Hills, C. A. Morrison, G. A. Turner, and M. R. Kokta, *Absorption Spectra and Energy Levels of Gd^{3+}, Nd^{3+}, and Cr^{3+} in the Garnet $Gd_3Sc_2Ga_3O_{12}$*, Phys. Rev. **B37** (1988), 8564.

N.F.M. Henry and K. Lonsdale, *International Tables for X-Ray Crystallography*, vol. I: *Symmetry Groups*, Kynoch, Birmingham, U.K. (1969).

T. Inui, Y. Tanabe, and Y. Onodera, *Group Theory and Its Applications in Physics*, Springer-Verlag, New York (1990).

B. R. Judd, *Operator Techniques in Atomic Spectroscopy*, McGraw-Hill, New York (1963), p 221.

N. Karayianis and C. A. Morrison, *Rare Earth Ion-Host Interactions: 1.—Point Charge Lattice Sum in Scheelites*, Harry Diamond Laboratories, HDL-TR-1648 (October 1973) (NTIS 011252).

E. König and S. Kramer, *Ligand Field Energy Diagrams*, Plenum Press, New York (1977).

R. P. Leavitt, *On the Role of Certain Rotational Invariants in Crystal-Field Theory*, J. Chem. Phys. **77** (1982), 1661.

A. D. Liehr, *The Three Electron (or Hole) Cubic Ligand Field Spectra*, J. Phys. Chem. **67** (1963), 1314. (This paper contains a tremendous number of references as well as considerable original analysis of experimental data. It also contains Tanabe-Sugano diagrams for the d^3 and d^7 electronic configuration for octahedral and tetrahedral symmetries including the spin-orbit interaction.)

R. M. Macfarlane, *Analysis of the Spectrum d^3 Ions in Trigonal Crystal Fields*, J. Chem. Phys. **39** (1963), 3118.

D. S. McClure, *Electronic Spectra of Molecules and Ions in Crystals: II*, Solid State Phys. **9** (1959), 420, Academic Press, New York.

C. A. Morrison, *Dipolar Contributions to the Crystal Fields in Ionic Solids*, Solid State Commun. **18** (1976), 153. (See also Faucher and Garcia, 1982, 1983; Garcia, Faucher, and Malta, 1983; and Garcia and Faucher, 1984.)

C. A. Morrison, *Host Dependence of the Rare-Earth Ion Energy Separation $4f^N - 4f^{N-1} nl$*, J. Chem. Phys. **72** (1980), 1001. In the expression for the Slater-parameter host-dependent shift, a factor $(k+1)$ was omitted. (See also Eremin and Kornienko, 1982.)

C. A. Morrison, *Angular Momentum Theory Applied to Interactions in Solids*, Lecture Notes in Chemistry **47**, Springer-Verlag, New York (1988).

C. A. Morrison, G. F. de Sá, and R. P. Leavitt, *Self-Induced Multipole Contribution to the Single-Electron Crystal Field*, J. Chem. Phys. **76** (1982), 3899.

C. A. Morrison and R. P. Leavitt, *Spectroscopic Properties of Triply Ionized Lanthanides in Transparent Host Crystals*, in *Handbook of the Physics and Chemistry of Rare Earths*, vol 5, K. Gschneidner and L. Eyring, eds., North-Holland, New York (1982).

C. A. Morrison and R. P. Leavitt, *Crystal Field Splittings of the Hund Ground States of nd^N Ions in S_4 Symmetry: Theory and Applications to the Ga^{3+} Site of $Gd_3Sc_2Ga_3O_{12}$*, Harry Diamond Laboratories, HDL-TR-2040 (March 1984).

C. A. Morrison and R. G. Schmalbach, *Approximate Values of $<r^k>$ for the Divalent, Trivalent, and Quadrivalent Ions with the $3d^N$ Electronic Configuration*, Harry Diamond Laboratories, HDL-TL-85-3 (July 1985).

C. W. Nielson and G. F. Koster, *Spectroscopic Coefficients for the p^n, d^n, and f^n Configurations*, MIT Press, Cambridge, MA (1963).

L. E. Orgel, *An Introduction to Transition-Metal Chemistry: Ligand-Field Theory*, John Wiley and Sons, New York (1960).

M.H.L. Pryce and W. A. Runciman, *The Absorption Spectrum of Vanadium in Corundum*, Discuss. Faraday Soc. **26** (1958), 34.

G. Racah, *Theory of Complex Spectra IV*, Phys. Rev. **76** (1949), 1352.

J. C. Slater, *Quantum Theory of Atomic Structure*, McGraw-Hill, New York (1960).

Y. Tanabe and S. Sugano, *On the Absorption Spectra of Complex Ions: I*, J. Phys. Soc. Japan **9** (1954), 753.

Y. Tanabe and S. Sugano, *On the Absorption Spectra of Complex Ions: II*, J. Phys. Soc. Japan **9** (1954), 766.

R. E. Trees, *$4f^3$ and $4f^2d$ Configuration of Doubly Ionized Praseodymium (Pr III)*, J. Opt. Soc. Am. **54** (1964), 651.

H. Watanabe, *Operator Methods in Ligand Field Theory*, Prentice-Hall, Englewood Cliffs, NJ (1966).

Table 1. Free-ion data: $F^{(2)}$, $F^{(4)}$, ζ, and α for nd^N ions (cm^{-1})
(A) $3d^N$

nd^N	Ion	$F^{(2)}$	$F^{(4)}$	ζ	α	Reference
$3d^1$	Sc^{2+}	—	—	79	—	7
$3d^1$	Ti^{3+}	—	—	158	—	2
$3d^1$	V^{4+}	—	—	253	—	6
$3d^2$	Ti^{2+}	53,061	30,920	126	56.4	3
$3d^2$	Ti^{2+}	54,870	32,034	129	20.80	5
$3d^2$	Ti^{2+}	54,927	32,206	118	20.52	9
$3d^2$	V^{3+}	67,200	40,522	220	75	4
$3d^2$	V^{3+}	69,547	42,234	206	27.54	9
$3d^2$	Cr^{4+}	75,831	47,061	338	—	4
$3d^2$	Cr^{4+}	82,406	50,755	319	37.64	9
$3d^3$	V^{2+}	55,153	20,954	186	199	4
$3d^3$	V^{2+}	59,669	35,882	177	24.58	5
$3d^3$	V^{2+}	59,924	36,268	170	22.90	9
$3d^3$	Cr^{3+}	75,950	30,076	296	437	4
$3d^3$	Cr^{3+}	74,201	45,822	275	29.87	9
$3d^3$	Mn^{4+}	80,332	47,754	437	91	4
$3d^3$	Mn^{4+}	86,939	54,219	405	39.01	9
$3d^4$	Cr^{2+}	62,300	38,934	263	61.0	4
$3d^4$	Cr^{2+}	64,467	39,730	239	28.36	5
$3d^4$	Cr^{2+}	64,798	40,288	231	25.83	9
$3d^4$	Mn^{3+}	81,970	46,998	388	12	4
$3d^4$	Mn^{3+}	78,756	49,404	—	32.60	9
$3d^4$	Fe^{4+}	87,269	56,183	565	85	4
$3d^4$	Fe^{4+}	91,372	57,696	513	40.66	9
$3d^5$	Mn^{2+}	67,685	40,698	351	74.8	3
$3d^5$	Mn^{2+}	69,266	43,578	317	32.14	5
$3d^5$	Mn^{2+}	69,485	44,305	316	29.20	9
$3d^5$	Fe^{3+}	83,302	53,070	463	35.40	9
$3d^5$	Co^{4+}	95,819	61,152	654	42.68	9
$3d^6$	Fe^{2+}	79,149	49,153	440	81	4
$3d^6$	Fe^{2+}	74,064	47,426	411	35.92	5
$3d^6$	Fe^{2+}	74,282	48,241	422	33.21	9
$3d^6$	Co^{3+}	87,762	56,823	606	37.93	9
$3d^6$	Ni^{4+}	100,186	64,788	830	44.17	9
$3d^7$	Co^{2+}	77,532	50,123	560	65	4
$3d^7$	Co^{2+}	78,906	52,277	536	37.48	9
$3d^7$	Co^{2+}	78,863	51,274	520	39.70	5
$3d^7$	Ni^{3+}	92,204	60,579	749	41.01	9
$3d^7$	Cu^{4+}	104,534	68,395	1008	46.40	9

Table 1 (cont'd). Free-ion data: $F^{(2)}$, $F^{(4)}$, ζ, and α for nd^N ions (cm^{-1})
(A) $3d^N$ (cont'd)

nd^N	Ion	$F^{(2)}$	$F^{(4)}$	ζ	α	Reference
$3d^8$	Ni^{2+}	86,933	60,871	702	42	4
$3d^8$	Ni^{2+}	83,661	55,122	644	43.48	5
$3d^8$	Ni^{2+}	83,514	56,164	668	42.49	9
$3d^8$	Cu^{3+}	96,631	64,302	911	44.79	9
$3d^8$	Zn^{4+}	108,877	71,954	1203	49.84	9
$3d^9$	Cu^{2+}	—	—	829	—	8
$3d^9$	Zn^{3+}	—	—	1155	—	10
$3d^9$	Ga^{4+}	—	—	1433	—	1

References for part A, $3d^N$ ions

1. E. Biemont and J. Hansen, *Energy Levels and Transition Probabilities in 3d and 3d^9 Configurations*, Phys. Scr. **39** (1989), 308.

2. C. Corliss and J. Sugar, *Energy Levels of Titanium, Ti I through Ti XXII*, J. Phys. Chem. Ref. Data **8** (1979), 1.

3. T. M. Dunn and W.-K. Li, *Magnetic Interactions for the Electronic Configuration d^5*, J. Chem. Phys. **46** (1967), 2907.

4. W.-K. Li, *Magnetic Interactions in Transition Metal Ions: Part I. Electronic Configurations d^2, d^3, and d^4*, Atomic Data **2** (1970), 45; *Part II. Bivalent Cations of the First Transition Series*, Atomic Data **2** (1970), 58.

5. A. Pasternak and Z. B. Goldschmidt, *Spin-Dependent Interactions in the 3d^N Configurations of the Third Spectra of the Iron Group*, Phys. Rev. **A6** (1972), 55. The parameters are given in the form:
$$F^{(2)} = 69{,}266 + 4798.5(N-5)$$
$$F^{(4)} = 43{,}578 + 3848(N-5)$$
$$\alpha = 32.14 + 3.78(N-5)$$
$$\zeta = 348.3 + 85.8(N-5) + 7.7[(N-5)^2 - 4]$$

6. J. Sugar and C. Corliss, *Energy Levels of Vanadium, V I through V XXIII*, J. Phys. Chem. Ref. Data **7** (1978), 1191.

7. J. Sugar and C. Corliss, *Energy Levels of Scandium: Sc I Through Sc XXI*, J. Phys. Chem. Ref. Data **9** (1980), 473.

8. A. G. Shenstone, *The Third Spectrum of Copper (Cu III)*, J. Res. Nat. Bur. Stand. Sect. A **79A** (1975), 497.

9. P.H.M. Uylings, A.J.J. Raassen, and J. F. Wyart, *Energies of N Equivalent Electrons Expressed in Terms of Two-Electron Energies and Independent Three-Electron Parameters: A New Complete Set of Orthogonal Operators: II. Application of 3d^N Configurations*, J. Phys. **B17** (1984), 4103

10. Th. A. M. Vankleff and Y. N. Joshi, *Spectrum of Quadruply Ionized Zinc: Zn V*, Phys. Rev. **25** (1982), 21017A.

Table 1 (cont'd). Free-ion data: $F^{(2)}$, $F^{(4)}$, ζ, and α for nd^N ions (cm^{-1})
(B) $4d^N$

$4d^N$	Ion	$F^{(2)}$	$F^{(4)}$	ζ	α	Reference
$4d^1$	Y^{2+}	—	—	290	—	11
$4d^1$	Zr^{3+}	—	—	500.	—	1
$4d^1$	Nb^{4+}	—	—	742	—	14
$4d^2$	Zr^{2+}	34,790	23,373	—	—	5
$4d^2$	Zr^{2+}	38,721	19,498	408	89.94	9
$4d^2$	Zr^{2+}	37,170	20,160	450	25	13
$4d^2$	Nb^{3+}	47,297	31,781	647	—	10
$4d^2$	Mo^{4+}	54,755	35,818	921	44.48	15
$4d^3$	Nb^{2+}	39,950	26,901	—	—	5
$4d^3$	Nb^{2+}	41,517	25,427	535	33	13
$4d^3$	Mo^{3+}	50,411	32,830	810	38	6
$4d^3$	Tc^{4+}	—	—	—	—	—
$4d^4$	Mo^{2+}	45,688	30,027	699	31	7
$4d^4$	Mo^{2+}	45,080	30,429	—	—	5
$4d^4$	Tc^{3+}	—	—	—	—	—
$4d^4$	Ru^{4+}	—	—	—	—	—
$4d^5$	Tc^{2+}	50,225	33,957	—	—	5
$4d^5$	Ru^{3+}	—	—	—	—	—
$4d^5$	Rh^{4+}	—	—	—	—	—
$4d^6$	Ru^{2+}	55,370	37,485	—	—	5
$4d^6$	Rh^{3+}	—	—	—	—	—
$4d^6$	Pd^{4+}	66,071	46,004	1853	24.8	12
$4d^7$	Rh^{2+}	60,515	41,013	—	—	5
$4d^7$	Rh^{2+}	54,117	38,582	1291	29	13
$4d^7$	Pd^{3+}	61,943	43,516	1699	31.6	2
$4d^7$	Ag^{4+}	71,497	51,108	2289	32.2	16
$4d^8$	Pd^{2+}	65,660	44,541	—	—	5
$4d^8$	Pd^{2+}	57,302	41,933	1545	28	13
$4d^8$	Pd^{2+}	57,766	42,591	1551	21.9	3
$4d^8$	Ag^{3+}	65,305	46,002	1996	45.98	17
$4d^8$	Cd^{4+}	72,155	50,707	2495	47.78	18
$4d^9$	Ag^{2+}	—	—	1844	—	4
$4d^9$	Ag^{2+}	—	—	1825	—	13
$4d^9$	Cd^{3+}	—	—	2325	—	18
$4d^9$	In^{4+}	—	—	2866	—	11

References for part B, $4d^N$ ions

1. N. Aquista and J. Reader, *Spectrum and Energy Levels of Triply Ionized Zirconium (Zr IV)*, J. Opt. Soc. Am. **70** (1980), 789.

20

2. M. M. Barakat, Th.A.M. Van Kleef, and A.J.J. Raassen, *Analysis of the Fourth Spectrum of Palladium: Pd IV 1. $4d^7$-$4d^6$ 5p Transitions*, Physica **132C** (1985), 240.

3. M. M. Barakat, Th.A.M. Van Kleef, and A.J.J. Raassen, *Extension of the Analysis of the Three Lowest Configurations in the Third Spectrum of Palladium (Pd III)*, Physica **132C** (1985), 111.

4. H. Benschop, Y. N. Joshi, and Th.A.M. Van Kleef, *The Spectrum of Doubly Ionized Silver: Ag III*, Can. J. Physics **53** (1975), 498.

5. L. DiSipio, E. Tondello, G. DeMichelis, and L. Oleari, *Slater-Condon Parameters for Atoms and Ions of the Second Transition Metal Series*, Inorg. Chem. **9** (1970), 927.

6. M. T. Ferdandez, I. Cabeza, L. Iglesias, O. Garcia-Riquelme, F. R. Rico, and V. Kaufman, *Fundamental Configurations in Mo IV Spectrum*, Phys. Scr. **35** (1987), 819.

7. L. Iglesias, M. I. Cabeza, and V. Kaufman, *Analysis of the Spectrum of Doubly Ionized Molybdenum (MoIII)*, J. Res. Nat. Inst. Stand. Technol. **95** (1990), 647.

8. Y. N. Joshi and Th.A.M. Van Kleef, *$4d^9$-$4d^8$ 5p Transitions in Cd IV, Sn VI, and Sb VII and Resonance Lines of Sn V and Sb VI*, Can. J. Phys. **55** (1977), 714.

9. Z. A. Khan, M. S. Z. Chaghtai, and K. Rahimullah, *Classified Lines and Energy Levels of Doubly Ionized Zirconium*, Phys. Scr. **23** (1981), 29.

10. E. Meinders, F. G. Meijer, and L. Remijin, *The Spectrum of Triply Ionized Niobium*, Phys. Scr. **25** (1982), 527.

11. C. Moore, *Atomic Energy Levels*, National Bureau of Standards Circular 467, vol 2–3, U.S. Government Printing Office, Washington, D.C. (1952–1958); reprinted 1971, NSRDS-NBS 35.

12. A.J.J. Raassen and Th.A.M. Van Kleef, *Analysis of the Fifth Spectrum of Palladium (Pd V)*, Physica **142c** (1986), 359.

13. Y. Shadmi, *The Configurations $4d^n$ + $4d^{n-1}$ 5s in Doubly-Ionized Atoms of the Palladium Group*, J. Res. Nat. Bur. Stand. Sect. A. **70A** (1966), 435.

14. Q. Shujauddin, A. Mushtaq, and M. S. Z. Chaghtai, *The Fifth Spectrum of Niobium (Nb V)*, Phys. Scr. **25** (1982), 924.

15. J. Sugar and A. Musgrove, *Energy Levels of Molybdenum, Mo I Through Mo XLII*, J. Phys. Chem. **17** (1988), 155.

16. Th.A.M. Van Kleef and Y. N. Joshi, *$4d^7$ 5s-$4d^7$ 5p Transitions in Ag IV*, Can. J. Phys. **61** (1983), 36.

17. Th.A.M. Van Kleef and Y. N. Joshi, *Analysis of $4d^8$-$4d^7$ 5p Transitions in Triply Ionized Silver: Ag IV*, Can. J. Phys. **59** (1981), 1930.

18. Th.A.M. Van Kleef, Y. N. Joshi, and R. P. Srivastava, *Analysis of Cd V: I.—$4d^8$-$4d^7$ 5p Transitions*, Physica **114C** (1982), 105.

Table 1 (cont'd). Free-ion data: $F^{(2)}$, $F^{(4)}$, ζ, and α for nd^N ions (cm^{-1})
(C) $5d^N$

$5d^N$	Ion	$F^{(2)}$	$F^{(4)}$	ζ	α	Reference
$5d^1$	Lu^{2+}	—	—	1176	—	6
$5d^1$	Hf^{3+}	—	—	1877	—	7
$5d^1$	Ta^{4+}	—	—	2643	—	9
$5d^2$	Hf^{2+}	—	—	—	—	—
$5d^2$	Ta^{3+}	45,551	28,658	2281	66.22	10
$5d^2$	W^{4+}	52,112	34,335	3102	—	8
$5d^3$	Ta^{2+}	—	—	—	—	—
$5d^3$	W^{3+}	47,530	29,988	2720	25	3
$5d^3$	Re^{4+}	—	—	—	—	—
$5d^4$	W^{2+}	43,442	28,186	2364	13	2
$5d^4$	Re^{3+}	—	—	—	—	—
$5d^4$	Os^{4+}	—	—	—	—	—
$5d^5$	Re^{2+}	—	—	—	—	—
$5d^5$	Os^{3+}	—	—	—	—	—
$5d^5$	Ir^{4+}	—	—	—	—	—
$5d^6$	Os^{2+}	—	—	—	—	—
$5d^6$	Ir^{3+}	—	—	—	—	—
$5d^6$	Pt^{4+}	—	—	—	—	—
$5d^7$	Ir^{2+}	—	—	—	—	—
$5d^7$	Pt^{3+}	—	—	—	—	—
$5d^7$	Au^{4+}	—	—	—	—	—
$5d^8$	Pt^{2+}	—	—	—	—	—
$5d^8$	Au^{3+}	55,462	42,089	5,539	56	5
$5d^8$	Hg^{4+}	—	—	—	—	—
$5d^9$	Au^{2+}	—	—	5078	—	1
$5d^9$	Hg^{3+}	—	—	6274	—	4
$5d^9$	Tl^{4+}	—	—	—	—	—

References for part C, $5d^N$ ions

1. J. C. Ehrhardt and S. P. Davis, *Precision Wavelengths and Energy Levels of Gold*, J. Opt. Soc. Am. **61** (1971), 1342.

2. L. Iglesias, M. Cabeza, F. Rico, O. Garcia-Riquelme, and V. Kaufman, *Spectrum of Doubly-Ionized Tungsten (W III)*, J. Res. Nat. Inst. Stand. Technol. **94** (1989), 221.

3. L. Iglesias, V. Kaufman, O. Garcia-Requelme, and F. R. Rico, *Analysis of the Fourth Spectrum of Tungsten (W IV)*, Phys. Scr. **31** (1985), 173.

4. Y. N. Joshi, A.J.J. Raassen, and B. Arcimowicz, *Fourth Spectrum of Mercury: Hg IV*, J. Opt. Soc. Am. **B6** (April 1989), 534–538.

5. Y. N. Joshi, A.J.J. Raassen, and A. A. van der Valk, *Analysis of the $5d^8$–$5d^768$ Transitions in Au IV*, Opt Soc. Am. 88 (1991), 1372.

6. V. Kaufman and J. Sugar, *One-Electron Spectrum of Doubly Ionized Lutetium (Lu III) and Nuclear Moment*, J. Opt. Soc. Am. **61** (1971), 1693.

7. P.F.A. Klinkenberg, Th.A.M. Van Kleef, and P. E. Noorman, *Spectral Structure of Doubly and Triply Ionized Hafnium*, Physica **27** (1961), 1177.

8. F. G. Meijer, *The Fifth Spectrum of Tungsten, W V*, Physica **141C** (1986), 230.

9. F. G. Meijer and P.F.A. Klinkenberg, *The Structure of the Spectrum of Ta V*, Physica (Utrecht) **69** (1973), 111.

10. F. G. Meijer and B. C. Metsch, *The Analysis of the Fourth Spectrum of Tantalum, Ta IV*, Physica (Utrecht) **94C** (1978), 259.

11. C. Moore, *Atomic Energy Levels*, National Bureau of Standards Circular 467, vol 3, U.S. Government Printing Office, Washington, D.C. (1952–1958); reprinted 1971, NSRDS-NBS 35.

Table 1D. Number of B_{nm} or A_{nm} for the d^N configuration for 30 of 32 point groups[a]

No.[b]	Group	Number of B_{nm} or A_{nm} for					Host	Site
		$n=1$	$n=2$	$n=3$	$n=4$	$n=5$		
3	C_2	1	3	3	5	5	Al_2O_3	O
4	C_s	2	3	4	5	6	$BeAl_2O_4$[c]	Al_2
5	C_{2h}	0	3	0	5	0	$La_{2-x}Sr_xCuO_4$	Cu
6	D_2	0	2	1	3	2	$A_3B_2Ge_3O_{12}$	A
7	C_{2v}	1	2	2	3	3	MnF_2	F
8	D_{2h}	0	2	0	3	0	ZnF_2	Zn
9	C_4	1	1	1	3	3	–	–
10	S_4	0	1	2	3	2	$Y_3Ga_5O_{12}$	Ga_1
11	C_{4h}	0	1	0	3	0	–	–
12	D_4	0	1	0	2	1	–	–
13	C_{4v}	1	1	1	2	2	$La_{2-x}Sr_xCuO_4$	La, Sr
14	D_{2d}	0	1	1	2	1	$LiMgZrO_4$	Mg, Zr
15	D_{4h}	0	1	0	2	0	$La_{2-x}Sr_xCuO_4$	Cu
16	C_3	1	1	3	3	3	Al_2O_3	Al
17	C_{3i}	0	1	0	3	0	$Y_3Al_5O_{12}$	Al_2
18	D_3	0	1	1	2	1	$Be_3Al_2(SiO_3)_6$	Al
19	C_{3v}	1	1	2	2	2	ZnO	Zn
20	D_{3d}	0	1	0	2	0	K_2ReF_2	Re
21	C_6	1	1	1	1	1	–	–
22	C_{3h}	0	1	2	1	2	–	–
23	C_{6h}	0	0	0	0	0	--	–
24	D_6	0	1	0	1	0	–	–
25	C_{6v}	1	1	1	1	1	–	–
26	D_{3h}	0	1	1	1	1	–	–
27	D_{6h}	0	1	0	1	0	–	–
28	T	0	0	1	1	0	$Bi_{12}GeO_{20}$[d]	Ge
29	T_h	0	0	0	1	0	–	–
30	O	0	0	0	1	0	–	–
31	T_d	0	0	1	1	0	$ZnGa_2O_4$	Zn
32	O_h	0	0	0	1	0	$KMgF_3$	Mg

[a]Morrison (1988).

[b]These numbers correspond with Koster et al (1963).

[c]Also an Al site with C_i symmetry.

[d]The Bi site has C_1 symmetry.

Table 2. Hartree-Fock values for $F^{(k)}$, ζ, and $<r^k>$ for nd^N ions

(A) $3d^N$

Z	X^{2+}	nd^N	$F^{(2)}$ (cm^{-1})	$F^{(4)}$ (cm^{-1})	ζ_d (cm^{-1})	$<r^2>$ (Å^2)	$<r^4>$ (Å^4)
21	Sc	$3d^1$	—	—	85.95	0.8346	1.4997
22	Ti	$3d^2$	67,932	42,357	131.15	0.6716	0.9808
23	V	$3d^3$	74,062	46,171	187.17	0.5677	0.7112
24	Cr	$3d^4$	79,790	49,726	256.60	0.4910	0.5401
25	Mn	$3d^5$	85,637	53,368	342.85	0.4277	0.4145
26	Fe	$3d^6$	89,877	55,927	441.38	0.3893	0.3527
27	Co	$3d^7$	94,600	58,817	561.21	0.3525	0.2949
28	Ni	$3d^8$	99,392	61,756	703.19	0.3203	0.2478
29	Cu	$3d^9$	—	—	869.65	0.2923	0.2097

Z	X^{3+}	nd^N	$F^{(2)}$ (cm^{-1})	$F^{(4)}$ (cm^{-1})	ζ_d (cm^{-1})	$<r^2>$ (Å^2)	$<r^4>$ (Å^4)
22	Ti	$3d^1$	—	—	157.75	0.5341	0.5769
23	V	$3d^2$	82,940	52,097	220.47	0.4571	0.4270
24	Cr	$3d^3$	88,514	55,558	296.26	0.4018	0.3344
25	Mn	$3d^4$	93,852	58,861	388.01	0.3578	0.2688
26	Fe	$3d^5$	99,367	62,291	499.53	0.3196	0.2168
27	Co	$3d^6$	103,474	64,758	625.26	0.2947	0.1884
28	Ni	$3d^7$	108,043	67,546	775.62	0.2705	0.1615
29	Cu	$3d^8$	112,696	70,392	951.32	0.2489	0.1389
30	Zn	$3d^9$	—	—	1154.85	0.2297	0.1200

Z	X^{4+}	nd^N	$F^{(2)}$ (cm^{-1})	$F^{(4)}$ (cm^{-1})	ζ_d (cm^{-1})	$<r^2>$ (Å^2)	$<r^4>$ (Å^4)
23	V	$3d^1$	—	—	253.27	0.4398	0.6494
24	Cr	$3d^2$	96,286	60,775	337.03	0.3708	0.4203
25	Mn	$3d^3$	101,615	64,078	436.20	0.3568	0.4097
26	Fe	$3d^4$	106,766	67,260	554.03	0.2914	0.2686
27	Co	$3d^5$	112,112	70,581	694.80	0.2721	0.2271
28	Ni	$3d^6$	116,164	73,011	851.77	0.2497	0.1875
29	Cu	$3d^7$	120,659	75,748	1036.93	0.2283	0.1524
30	Zn	$3d^8$	125,241	78,548	1250.84	0.2100	0.1200
31	Ga	$3d^9$	—	—	1496.12	0.1800	0.0800

S. Fraga, K.M.S. Saxena, and J. Karwowski, *Handbook of Atomic Data*, Elsevier, New York (1976).

Table 2 (cont'd). Hartree-Fock values for $F^{(k)}$, ζ, and $<r^k>$ for nd^N ions
(B) $4d^N$

Z	X^{2+}	nd^N	$F^{(2)}$ (cm^{-1})	$F^{(4)}$ (cm^{-1})	ζ_d (cm^{-1})	$<r^2>$ (Å^2)	$<r^4>$ (Å^4)
39	Y	$4d^1$	—	—	312	1.5737	4.4402
40	Zr	$4d^2$	51,177	33,321	432	1.2734	2.8974
41	Nb	$4d^3$	55,683	36,328	566	1.0769	2.0761
42	Mo	$4d^4$	59,873	39,117	718	0.9316	1.5580
43	Tc	$4d^5$	64,052	41,911	891	0.8145	1.1907
44	Ru	$4d^6$	67,247	43,978	1082	0.7365	0.9869
45	Rh	$4d^7$	70,673	46,224	1299	0.6656	0.8126
46	Pd	$4d^8$	74,108	48,480	1544	0.6045	0.6744
47	Ag	$4d^9$	—	—	1820	0.5516	0.5644

Z	X^{3+}	nd^N	$F^{(2)}$ (cm^{-1})	$F^{(4)}$ (cm^{-1})	ζ_d (cm^{-1})	$<r^2>$ (Å^2)	$<r^4>$ (Å^4)
40	Zr	$4d^1$	—	—	510	1.0840	1.989
41	Nb	$4d^2$	60,253	36,327	655	0.9288	1.461
42	Mo	$4d^3$	64,276	42,326	815	0.8149	1.128
43	Tc	$4d^4$	68,116	44,878	997	0.7244	0.8953
44	Ru	$4d^5$	72,001	47,470	1201	0.6479	0.7175
45	Rh	$4d^6$	75,061	49,443	1426	0.5936	0.6094
46	Pd	$4d^7$	78,342	51,586	1680	0.5435	0.5147
47	Ag	$4d^8$	81,645	53,754	1964	0.4993	0.4372
48	Cd	$4d^9$	—	—	2283	0.4603	0.3736

Z	X^{4+}	nd^N	$F^{(2)}$ (cm^{-1})	$F^{(4)}$ (cm^{-1})	ζ_d (cm^{-1})	$<r^2>$ (Å^2)	$<r^4>$ (Å^4)
41	Nb	$4d^1$	—	—	742	0.8715	1.559
42	Mo	$4d^2$	68,068	45,102	914	0.7520	1.085
43	Tc	$4d^3$	71,843	47,615	1105	0.7039	0.9993
44	Ru	$4d^4$	75,495	50,038	1319	0.5959	0.6777
45	Rh	$4d^5$	79,206	52,512	1557	0.5522	0.5776
46	Pd	$4d^6$	82,196	54,434	1820	0.5071	0.4823
47	Ag	$4d^7$	85,389	56,519	2114	0.4654	0.4005
48	Cd	$4d^8$	88,615	58,631	2441	0.4200	0.3200
49	In	$4d^9$	—	—	2806	0.3800	0.2400

S. Fraga, K.M.S. Saxena, and J. Karwowski, *Handbook of Atomic Data*, Elsevier, New York (1976).

26

Table 2 (cont'd). Hartree-Fock values for $F^{(k)}$, ζ, and $<r^k>$ for nd^N ions
(C) $5d^N$

Z	X^{2+}	nd^N	$F^{(2)}$ (cm^{-1})	$F^{(4)}$ (cm^{-1})	ζ_d (cm^{-1})	$<r^2>$ (Å^2)	$<r^4>$ (Å^4)
71	Lu	$5d^1$	—	—	1391	1.6197	4.6324
72	Hf	$5d^2$	50,350	33,000	1774	1.3646	3.2437
73	Ta	$5d^3$	54,008	35,526	2170	1.1926	2.4612
74	W	$5d^4$	57,369	37,840	2594	1.0610	1.9385
75	Re	$5d^5$	60,702	40,149	3053	0.9510	1.5467
76	Os	$5d^6$	63,123	41,766	3531	0.8779	1.3277
77	Ir	$5d^7$	65,755	43,550	4056	0.8087	1.1289
78	Pt	$5d^8$	68,388	45,344	4626	0.7474	0.9649
79	Au	$5d^9$	—	—	5248	0.6930	0.6646

Z	X^{3+}	nd^N	$F^{(2)}$ (cm^{-1})	$F^{(4)}$ (cm^{-1})	ζ_d (cm^{-1})	$<r^2>$ (Å^2)	$<r^4>$ (Å^4)
72	Hf	$5d^1$	—	—	2072	1.1760	2.2810
73	Ta	$5d^2$	58,176	38,604	2489	1.0420	1.7780
74	W	$5d^3$	61,293	40,754	2926	0.9399	1.4440
75	Re	$5d^4$	64,247	42,789	3394	0.8563	1.1960
76	Os	$5d^5$	67,234	44,856	3898	0.7829	0.9968
77	Ir	$5d^6$	69,479	46,349	4430	0.7309	0.8745
78	Pt	$5d^7$	71,922	48,001	5011	0.6809	0.7612
79	Au	$5d^8$	74,389	49,674	5640	0.6356	0.6646
80	Hg	$5d^9$	—	—	6323	0.5948	0.5826

Z	X^{4+}	nd^N	$F^{(2)}$ (cm^{-1})	$F^{(4)}$ (cm^{-1})	ζ_d (cm^{-1})	$<r^2>$ (Å^2)	$<r^4>$ (Å^4)
73	Ta	$5d^1$	—	—	2797	0.9823	1.8522
74	W	$5d^2$	64,633	43,252	3257	0.8727	1.3654
75	Re	$5d^3$	67,469	45,208	3741	0.8339	1.2966
76	Os	$5d^4$	70,197	47,082	4259	0.7247	0.9217
77	Ir	$5d^5$	72,980	49,004	4814	0.6841	0.8141
78	Pt	$5d^6$	75,122	50,424	5404	0.6393	0.7024
79	Au	$5d^7$	77,448	51,994	6045	0.5969	0.6030
80	Hg	$5d^8$	79,808	53,587	6736	0.5500	0.5000
81	Tl	$5d^9$	—	—	7486	0.5100	0.4000

S. Fraga, K.M.S. Saxena, and J. Karwowski, *Handbook of Atomic Data*, Elsevier, New York (1976).

3. $Y_3Al_5O_{12}$ (YAG)

3.1 Crystallographic Data on $Y_3Al_5O_{12}$

Cubic O_h^{10} (Ia3d), 230, Z = 8

Ion	Site	Symmetry	x^a	y	z	q	$\alpha\,(\text{Å}^3)^b$
Al_1	16(a)	C_{3i}	0	0	0	3	0.0530
Al_2	24(d)	S_4	0	1/4	3/8	3	0.0530
Y	24(c)	D_2	0	1/4	1/8	3	0.870
O	96(h)	C_1	−0.0306	0.0512	0.1500	−2	1.349

[a] X-ray data: a = 12.000 Å (Euler and Bruce, 1965).
[b] Schmidt et al (1979).

3.2 Crystal Fields for 24(d) (S_4) Site

3.2.1 Crystal-field components, A_{nm} (cm^{-1}/Å^n), for Al_2 (S_4) site

A_{nm}	Point charge	Self-induced	Dipole	Total
A_{20}	6,355	−2,604	14,013	17,765
ReA_{32}	−27,522	8,609	−11,957	−30,870
ImA_{32}	37,839	−11,913	6,332	32,258
A_{40}	−25,089	11,879	−8,516	−21,726
ReA_{44}	−3,763	1,614	1,964	−185.1
ImA_{44}	−9,108	4,740	−2,875	−7,243
ReA_{52}	−2,931	2,287	−3,498	−4,142
ImA_{52}	4,328	−3,207	3,640	4,762
$|A_{44}|$	9,855	—	—	7,245

[a] Morrison and Turner (1988).

3.2.2 Approximate Slater parameters (cm^{-1}) and crystal-field parameters (cm^{-1}) for triply ionized transition-metal ions in S_4 site in YAG[a]

X^{3+}	$3d^N$	$F^{(2)}$	$F^{(4)}$	B_{20}	B_{40}	B_{43}
Ti	d^1	—	—	2733	−33,697	13,236
V	d^2	56,732	36,132	2696	−32,714	12,850
Cr	d^3	59,983	38,912	2632	−31,730	12,464
Mn	d^4	63,234	41,688	2574	−30,747	12,077
Fe	d^5	66,485	44,463	2518	−29,766	11,692
Co	d^6	69,736	47,239	2450	−28,782	11,306
Ni	d^7	72,987	50,015	2387	−27,799	10,919
Cu	d^8	76,258	52,790	2326	−26,818	10,534
Zn	d^9	—	—	2267	−25,834	10,148

[a] Morrison and Turner (1988).

3.3 Crystal Fields for 16(a) (C_{3i}) Site

3.3.1 Crystal-field components, A_{nm} (cm^{-1}/Å^n), for Al$_1$ (C_{3i}) site (rotated so that z-axis is parallel to (111) crystallographic axis)

A_{nm}	Point charge	Self-induced	Dipole	Total
A_{20}	6,836	−1107	−13,553	−7,823
A_{40}	−20,054	8166	3,273	−8,615
ReA_{43}	2,813	−1422	6,253	7,644
ImA_{43}	−22,370	8639	2,348	−11,383
$|A_{43}|$	22,546	—	—	13,711

3.3.2 Slater parameters and theoretical crystal-field parameters, B_{nm} (cm^{-1}), for Al$_1$ (C_{3i}) site for triply ionized transition-metal ions with electronic configuration $3d^{Na}$

X^{3+}	$3d^N$	$F^{(2)}$	$F^{(4)}$	B_{20}	B_{40}	B_{43}
Ti	d^1	—	—	2953	−26,934	30,280
V	d^2	56,732	36,132	2895	−26,148	29,397
Cr	d^3	59,983	38,912	2832	−25,363	28,514
Mn	d^4	63,234	41,688	2769	−24,577	27,631
Fe	d^5	66,485	44,463	2709	−23,792	26,748
Co	d^6	69,736	47,239	2635	−26,006	25,865
Ni	d^7	72,987	50,015	2568	−22,221	24,982
Cu	d^8	76,238	52,790	2502	−21,435	24,099
Zn	d^9	—	—	2438	−20,650	23,216

[a]Morrison and Turner (1988).

3.4 Experimental Values (cm^{-1}) of B_{40}, $F^{(2)}$, and $F^{(4)}$ for nd^N Ions[a]

Ion	nd^N	$F^{(2)}$	$F^{(4)}$	B_{40}	Temp (K)	Site	Reference
Cr^{3+}	$3d^3$	53,438	34,978	−23,730	300	C_{3i}	1,3
Cr^{3+}	$3d^3$	55,776	36,806	−23,072	300	C_{3i}	28
Cr^{3+}	$3d^3$	53,438	40,320	−24,150	—	C_{3i}	26
Cr^{3+}	$3d^3$	—	—	−35,070	—	C_{3i}	22
Cr^{3+}	$3d^3$	54,600	40,950	−22,960	77	C_{3i}	35
Mn^{3+}	$3d^4$	59,500	32,130	−27,650	300	C_{3i}	1
Mn^{4+}	$3d^3$	53,543	43,457	−27,874	300	C_{3i}	28
Mn^{4+}	$3d^3$	—	—	−29,400	—	C_{3i}	2
Fe^{3+}	$3d^5$	39,690	15,876	−26,950	300	C_{3i}	1
Fe^{3+}	$3d^5$	49,224	36,477	−17,682	—	C_{3i}	21,30
Fe^{3+}	$3d^5$	51,023	42,979	−21,756	—	S_4	21,30
Fe^{3+}	$3d^5$	44,821	42,399	−16,904	—	S_4	23

[a]All cubic approximation. For C_{3i}, $B_{43} = \sqrt{10/7}\ B_{40}$. For S_4, $B_{44} = \sqrt{5/14}\ B_{40}$.

(cont'd)

Ion	nd^N	$F^{(2)}$	$F^{(4)}$	B_{40}	Temp (K)	Site	Reference
Co^{3+}	$3d^6$	56,630	34,020	−25,200	300	C_{3i}	1
Co^{3+}	$3d^6$	—	—	−17,430	—	S_4	36
Co^{2+}	$3d^7$	—	—	−9,660	—	S_4	36
Co^{2+}	$3d^7$	—	—	−12,880	—	C_{3i}	36
Ni^{3+}	$3d^7$	42,000	22,680	−27,580	300	C_{3i}	1,26
Ni^{2+}	$3d^8$	62,370	42,323	−10,815	300	S_4	8
Rh^{3+}	$4d^6$	40,600	21,924	−28,840	—	C_{3i}	21
Pd^{3+}	$4d^7$	39,326	21,218	−23,730	—	C_{3i}	21
Pt^{3+}	$5d^7$	44,520	30,744	−23,100	—	C_{3i}	21
V^{3+}	$3d^2$	—	—	−23,800	—	C_{3i}	21
V^{3+}	$3d^2$	—	—	−17,850	—	S_4	21
V^{4+}	$3d^1$	—	—	−30,800	—	C_{3i}	21

3.5   Bibliography and References

1. P. A. Arsenev and D. T. Sviridov, *Absorption Spectra of Yttrium Alumi-num Garnet (YAG) with Contaminant Ions of the Iron Group*, Sov. Phys. Crystallogr. **14** (1970), 578.

2. P. A. Arsenev, D. T. Sviridov, and N. P. Fialkovskaya, *Absorption Spectra of Aluminum-Yttrium Garnet Single Crystals Containing Palladium and Rhodium*, Sov. Phys. Crystallogr. **15** (1971), 711.

3. Z. T. Azamatov, P. A. Arsenev, T. Yu. Geraskina, and M. V. Chukichev, *Properties of Chromium Ions in the Lattice of Yttrium Aluminum Garnet (YAG)*, Phys. Status Solidi **(a)1** (1970), 801.

4. Kh. S. Bagdasarov, L. V. Bershov, V. O. Martirosyan, and M. L. Meilman, *The State of Molybdenum Impurity in Yttrium-Aluminum Garnet*, Phys. Status Solidi **(b)46** (1971), 745.

5. W. L. Bond, *Measurement of the Refractive Indices of Several Crystals*, J. Appl. Phys. **36** (1965), 1674.

6. G. Burns, E. A. Geiss, B. A. Jenkins, and M. I. Nathan, *Cr^{3+} Fluorescence in Garnets and Other Crystals*, Phys. Rev. **A139** (1965), 1687.

7. T. S. Chernaya, L. A. Muradyan, A. A. Rusakov, A. A. Kaminskii, and V. I. Simonov, *Refinement and Analysis of Atomic Structures of Er$_3$Al$_5$O$_{12}$ and (Y$_{2.80}$Er$_{0.20}$)Al$_5$O$_{12}$*, Sov. Phys. Crystallogr. **30** (1985), 38.

8. M. I. Demchuk, E. P. Dubrevina, N. V. Kullshov, V. P. Michailov, and V. A. Sandulenko, *Spectroscopy of Rare-Earth Garnet Crystals Activated by Nickel*, Opt. Spektrosk. **69** (1990) 1059.

9. D. P. Devor, R. C. Pastor, and L. G. DeShazer, *Hydroxyl Impurity Effects in YAG (Y$_3$Al$_5$O$_{12}$)*, J. Chem. Phys. **81** (1984), 4104.

10. I. N. Douglas, *Optical Spectra of Chromium Ions in Crystals of Yttrium Aluminum Garnet,* Phys. Status Solidi (a)**9** (1972), 635.

11. F. Euler and J. A. Bruce, *Oxygen Coordinates of Compounds with Garnet Structure,* Acta Crystallogr. **19** (1965), 971.

12. J. B. Gruber, M. E. Hills, C. A. Morrison, G. A. Turner, and M. R. Kokta, *Absorption Spectra and Energy Levels of Gd^{3+}, Nd^{3+}, and Cr^{3+} in the Garnet $Gd_3Sc_2Ga_3O_{12}$,* Phys. Rev. **B37** (1988), 8564.

13. J. A. Hodges, R. A. Serway, and S. A. Marshall, *Electron-Spin Resonance Absorption Spectrum of Platinum in Yttrium Aluminum Garnet,* Phys. Rev. **151** (1966), 196.

14. J. P. Hurrell, S.P.S. Porto, I. F. Chang, S. S. Mitra, and R. P. Bauman, *Optical Phonons of Yttrium Aluminum Garnet,* Phys. Rev. **173** (1968), 851.

15. C. Z. Janusz, W. Jelenski, and A. Niklas, *Disclosure of Defects in YAG:Nd Crystals by Thermoluminescence Method,* J. Cryst. Growth **57** (1982), 593.

16. I. I. Karpov, Kh. S. Bagdasarov, B. N. Grechushnikov, E. V. Antonov, and L. S. Garashina, *Conditions of Growth of Yttrium-Aluminum Garnet Crystals with Added Titanium,* Sov. Phys. Crystallogr. **23** (1978), 710.

17. I. I. Karpov, B. N. Grechushnikov, and Kh. S. Bagdasarov, *Color Centers in Titanium-Activated Yttrium-Aluminum Garnet Crystals,* Sov. Phys. Crystallogr. **23** (1978), 609.

18. I. I. Karpov, B. N. Grechushnikov, V. F. Koryagin, A. M. Kevorkov, and P. Z. Ngi, *EPR Spectra of V^{2+} Ions in Yttrium-Aluminum Garnet Crystals,* Sov. Phys. Dokl. **24** (1979), 33.

19. I. I. Karpov, B. N. Grechushnikov, V. F. Koryagin, A. M. Kevorkov, and P. Z. Ngi, *Investigation of V^{4+}-Ion Impurity in Crystals of Yttrium-Aluminum Garnet,* Sov. Phys. Dokl. **23** (1978), 492.

20. N. A. Kulagin, M. F. Ozerov, and V. O. Rokhmanova, *Effect of γ Radiation on the Electron State of Chromium Ions in $Y_3Al_5O_{12}$ Monocrystals,* Zh. Prikl. Spektrosk. (trans.) **46** (1987), 393.

21. Landolt-Bornstein, *Numerical Data and Functional Relationships in Science and Technology,* New Series, vol 12, Supplement and Extension to vol 4, part a, *Garnets and Perovskites,* Springer-Verlag, New York (1978), p 301.

22. R. W. McMillan, *Optical Absorption Spectrum of Cr^{3+} in Yttrium Aluminum Garnet,* J. Opt. Soc. Am. **67** (1977), 27.

23. M. L. Meil'man, M. V. Korzhik, V. V. Kuz'min, M. G. Livshits, Kh. S. Bagdasarov, and A. M. Kevorkov, *Luminescence and Energy-Level Structure of Impurity Centers in $Y_3Al_5O_{12}$: Fe^{3+} Single Crystals,* Sov. Phys. Dokl. **29** (1984), C1.

24. C. A. Morrison and G. A. Turner, *Analysis of the Optical Spectra of Triply Ionized Transition Metal Ions in Yttrium Aluminum Garnet (YAG),* Harry Diamond Laboratories, HDL-TR-2150 (October 1988).

25. P. C. Schmidt, A. Weiss, and T. P. Das, *Effect of Crystal Fields and Self-Consistency on Dipole and Quadrupole Polarizabilities of Closed-Shell Ions,* Phys. Rev. **B19** (1979), 5525.

26. B. K. Sevast'yanov, D. T. Sviridov, V. P. Orekhovo, L. B. Pasternak, R. K. Sviridova, and T. F. Vermeichik, *Optical Absorption Spectra of Excited Cr^{3+} Ions in Yttrium Aluminum Garnet,* Sov. J. Electron. **2** (1973), 339.

27. G. A. Slack, D. W. Oliver, R. M. Chrenko, and S. Roberts, *Optical Absorption of Y$_3$Al$_5$O$_{12}$ from 10- to 55000-cm^{-1} Wave Numbers,* Phys. Rev. **177** (1969), 1308.

28. D. T. Sviridov, R. K. Sviridova, N. I. Kulik, and V. B. Glasko, *Optical Spectra of the Iso-electronic Ions V^{2+}, Cr^{3+} and Mn^{4+} in an Octahedral Coordination,* J. Appl. Spectrosc. **30** (1979), 334.

29. W. F. van der Weg, Th. J. A. Popma, and A. T. Vink, *Concentration Dependence of UV and Electron-Excited Tb^{3+} Luminescence in Y$_3$Al$_5$O$_{12}$,* J. Appl. Phys. **57** (1985), 5450.

30. T. F. Veremeichik, B. N. Grechushnikov, T. M. Varina, D. T. Sviridov, and I. N. Kalinkina, *Absorption Spectra and Calculation of Energy-Level Diagram of Fe^{3+} and Mn^{2+} Ions in Single Crystals of Yttrium Aluminum Garnet, Orthoclase, and Manganese Silicate,* Sov. Phys. Crystallogr. **19** (1975), 742.

31. Yu. A. Voitukevich, M. V. Korzhik, V. V. Kuz'min, M. G. Livshits, and M. L. Meil'man, *Energy Structure of Iron (3$^+$) Impurity Ions in Yttrium Aluminum Garnet (Y$_3$Al$_5$O$_{12}$) Crystals,* Opt. Spectrosc. **63** (1987), 810.

32. R. Wannemacher and J. Heber, *Cooperative Emission of Photons by Weakly Coupled Chromium Ions in YAG and LaAlO$_3$,* J. Lumin. **39** (1987), 49.

33. M. J. Weber and L. A. Riseberg, *Optical Spectra of Vanadium Ions in Yttrium Aluminum Garnet,* J. Chem. Phys. **55** (1971), 2032.

34. M. Wojcik, *Estimation of Applicability of the Electrostatic Model to Calculate the Crystal Field Parameters in Garnets,* Physica B (Netherlands) **121B** (1983), 370.

35. D. L. Wood, J. Ferguson, K. Knox, and J. F. Dillon, Jr., *Crystal-Field Spectra of d^{3,7} Ions: III.—Spectrum of Cr^{3+} in Various Octahedral Crystal Fields,* J. Chem. Phys. **39** (1963), 890.

36. D. L. Wood and J. P. Remeika, *Optical Absorption of Tetrahedral Co^{3+} and Co^{2+} in Garnets,* J. Chem. Phys. **46** (1967), 3595.

37. C. A. Morrison, J. D. Bruno, and G. A. Turner, *Analysis of the Spectra of Triply Ionized Iron in Rare-Earth Aluminum Garnet,* Harry Diamond Laboratories, HDL-TR-2113 (September 1987).

38. C. A. Morrison and R. P. Leavitt, *Spectroscopic Properties of Triply Ionized Lanthanides in Transparent Host Materials,* in vol 5, *Handbook of the Physics and Chemistry of Rare Earths,* K. A. Geshneider, Jr., and L. Eyring, eds., North-Holland Publishers, New York (1982).

4. K_2NaXF_6 (X = Al, Ga, Sc)

4.1 Crystallographic Data on Two Forms of K_2NaAlF_6

4.1.1 Cubic T_h^6 ($Pa3$), 205, $Z = 4$, elpasoite

Ion	Site	Symmetry	x^a	y	z	q	α $(\text{Å}^3)^b$
Al	4(a)	C_{3i}	0	0	0	3	0.0530
Na	4(b)	C_{3i}	1/2	1/2	1/2	1	0.147
K	8(c)	C_3	1/4	1/4	1/4	1	0.827
F	24(d)	C_1	0.22	0.03	0.01	−1	0.731

[a]X-ray data: $a = 8.11$ Å (Wyckoff, 1988).
[b]Schmidt et al (1979).

4.1.2 Cubic O_h^5 ($Fm3m$), 225, $Z = 4$, elpasolite

Ion	Site	Symmetry	x	y	z	q	α $(\text{Å}^3)^a$
X	4(a)	O_h	0	0	0	3	α_x
Na	4(b)	O_h	1/2	1/2	1/2	1	0.147
K	8(c)	T_d	1/4	1/4	1/4	1	0.827
F	24(e)	C_{4v}	x	0	0	−1	0.731

[a]Schmidt et al (1979).

4.1.3 X-ray data for O_h^5 crystal structure

X	a	x_F	α_x $(\text{Å}^3)^b$	Ref
Al	8.119	0.219	0.0530	20
Ga	8.246	0.225^a	0.458	17,6
Sc	8.4717	0.2342	0.0540	22

[a]Private communication, R. H. Bartram (1990).
[b]Schmidt et al (1979).

4.2 Crystal Fields for $Pa3$ Form

4.2.1 Crystal-field components, A_{nm} (cm^{-1}/Å), for Al (C_{3i}) site (rotated so that z-axis is parallel to (111) crystallographic axis)[a]

A_{nm}	Monopole	Self-induced	Dipole	Total
A_{20}	20,455	−2717	4,145	21,883
A_{40}	−15,791	8018	−10,411	−18,186
ReA_{43}	14,510	−7295	9,008	17,095
ImA_{43}	3,769	−1945	7,546	4,370
$|A_{43}|$	14,992	—	—	17,645

[a]If the crystal-field components are computed in this crystal form, but with $y_F = z_F = 0$ and with a and x_F from table 4.1.3 for K_2NaAlF_6, then the resulting crystal-field components are indentical to the result given in the first row of table 4.3.1.

4.2.2 **Theoretical crystal-field parameters, B_{nm} (cm^{-1}), for monopole A_{nm} for Al (C_{3i}) site for triply ionized transition-metal ions with electronic configuration $3d^N$**

X^{3+}	N	B_{20}	B_{40}	B_{43}
Ti	1	16,285	−20,228	19,202
V	2	13,688	−14,441	13,708
Cr	3	11,820	−10,913	10,360
Mn	4	10,343	−8,469	8,039
Fe	5	9,078	−6,596	6,261
Co	6	8,227	−5,536	5,256
Ni	7	7,422	−4,584	4,352
Cu	8	6,712	−3,810	3,617
Zn	9	6,092	−3,184	3,022

4.3 Crystal Fields for *Fm3m* Form

4.3.1 **Crystal-field components, A_{nm} (cm^{-1}/Å^n), for Al (O_h) site**

X	A_{nm}[a]	Monopole	Self-induced	Dipole	Total	X-ray ref
Al	A_{40}	23,267	−12,274	15,593	26,586	20
Ga	A_{40}	18,871	−8,725	11,463	21,609	6,17
Sc	A_{40}	13,579	−5,093	7,043	15,520	22

[a]$A_{44} = \sqrt{5/14}\ A_{40}$.

4.3.2 **Experimental parameters for ions in O_h site in K$_2$NaXF$_6$**

Ion	X	$F^{(2)}$	$F^{(4)}$	ζ	B_{40}[a]	Ref
Cr^{3+}	Ga	58,380	38,052	—	33,810	13,18
Cr^{3+}	Sc	53,729	31,405	—	32,760	2[b]
Cr^{3+}	Ga	59,640	40,320	225	34,020	25
Cr^{3+}	Sc	61,516	40,698	—	32,760	2
Cr^{3+}	Ga	—	—	—	28,770	10
Cr^{3+}	Ga	—	—	—	31,080[c]	10
Cr^{3+}	Ga	60,986	41,773	—	33,600	3
Cr^{3+}	Sc	61,527	40,718	—	32,760	3
V^{3+}	Sc	54,761	35,154	124	32,718	22

[a]$B_{44} = \sqrt{5/14}\ B_{40}$.
[b]Experimental data of reference 2 were best fit by varying $F^{(2)}$ and $F^{(4)}$. B_{40} is their value.
[c]At 61 kbar pressure.

4.4   Bibliography and References

1. C. D. Adam, *ENDOR Determination of Covalency in K_2NaAlF_6, Cr^{3+}*, J. Phys. **C14** (1981), L105.

2. L. J. Andrews, S. M. Hitelman, M. Kokta, and D. Gabbe, *Excited State Absorption of Cr^{3+} in K_2NaScF_6 and $Gd_3Ga_2(MO_4)_3$, M = Ga, Al,* J. Chem. Phys. **84** (1986), 5229.

3. L. J. Andrews, A. Lempicki, B. C. McCollum, C. J. Giunta, R. H. Bartram, and J. F. Dolan, *Thermal Quenching of Chromium Photoluminescence in Ordered Perovskites. I. Temperature Dependence of Spectra and Lifetimes,* Phys. Rev. **B34** (1986) 2735.

4. L. J. Andrews, B. C. McCollum, and R. H. Bartram, *Cr^{3+} in K_2NaScF_6: Crystal Growth and Spectroscopy,* AIP Conf. Proc. **146** (1986), 227.

5. D. Babel, R. Haegele, G. Pausewang, and F. Wall, *Uber Kubische und Hexagonale Elpasolithe $A_2^I B^I M^{III} F_6$,* Mater. Res. Bull. **8** (1973), 1371 (in German).

6. R. H. Bartram, private communication (1990).

7. R. H. Bartram, J. C. Charpie, L. J. Andrews, and A. Lempicki, *Thermal Quenching of Chromium Photoluminescence in Ordered Perovskites. II. Theoretical Models,* Phys. Rev. **B34** (1986), 2741.

8. R. H. Bartram, J. F. Dolan, J. C. Charpie, A. G. Rinzler, and L. A. Kappers, *Pressure Dependence and Thermal Quenching of Chromium Photoluminescence in Ordered Perovskites,* Cryst. Latt. Def. Amorph. Mater. **15** (1987), 165.

9. J. M. Dance, J. Grannec, and A. Tressaud, *ESR Study of Nickel (+III) and Copper (+III) in Fluoride with Cryolite or Elpasolite-Type Structure,* Eur. J. Solid State Inorg. Chem. **25** (1988), 621.

10. J. F. Dolan, L. A. Kappers and R. H. Bartram, *Pressure and Temperature Dependence of Chromium Photoluminescence in K_2NaGaF_6,* Phys. Rev. **B33** (1986), 7339.

11. C. J. Donnelly, S. M. Healy, T. J. Glynn, G. F. Imbusch, and G. P. Morgan, *Spectroscopic Effects of Small $^4T_2$-^2E Energy Separations in $3d^3$-Ion Systems,* J. Lumin. **42** (1988), 119.

12. L. Dubicki, J. Ferguson, and B. van Oosterhout, *A Magneto-Optical Study of the $^2T_{1g}$, $^4T_{2g}$, $^2E_g \leftarrow ^4A_{2g}$ Transitions in K_2NaGaF_6:Cr^{3+},* J. Phys. **C13** (1980), 2791.

13. J. Ferguson, H. J. Guggenheim, and D. L. Wood, *Crystal-Field Spectra of $d^{3,7}$ Ions. VII. Cr^{3+} in K_2NaGaF_6,* J. Chem. Phys. **54** (1971), 504.

14. P. Greenough and A. G. Paulusz, *The $^2E_g \rightarrow ^4A_{2g}$ Phosphorescence Spectrum of the Cr^{3+} Ion in K_2NaAlF_6,* J. Chem. Phys. **70** (1979), 1967.

15. K. Grjotheim, J. L. Holm, and S. A. Mikhaiel, *Equilibrium Studies in the Systems K$_3$AlF$_6$-Na$_3$AlF$_6$ and K$_3$AlF$_6$-Rb$_3$AlF$_6$*, Acta Chem. Scand. **27** (1973), 1299.

16. P. T. Kenyon, L. Andrews, B. McCollum, and A. Lempicki, *Tunable Infrared Solid-State Laser Materials Based on Cr^{3+} in Low Ligand Fields,* IEEE J. Quant. Electron **QE18** (1982), 1189.

17. K. Knox and D. W. Mitchell, *The Preparation and Structure of K$_2$NaCrF$_6$, K$_2$NaFeF$_6$ and K$_2$NaGaF$_6$,* J. Inorg. Nucl. Chem. **21** (1961), 253.

18. A. D. Liehr, *The Three Electron (or Hole) Cubic Ligand Field Spectrum,* J. Phys. Chem. **67** (1963), 1314.

19. W. Massa, D. Babel, M. Epple, and W. Rudorff, *Are Fluoro Elpasolites Disordered? Structure Determinations of Single Crystals of K$_2$NaCrF$_6$, Rb$_2$NaFeF$_6$, and Rb$_2$KFeF$_6$,* Rev. Chim. Miner. **23** (1986), 508 (in German).

20. L. R. Morss, *Crystal Structure of Dipotassium Sodium Fluoroaluminate (Elpasolite),* J. Inorg. Nucl. Chem. **36** (1974), 3876.

21. S. A. Payne, L. L. Chase, and W. F. Krupke, *Optical Properties of Cr^{3+} in Fluorite-Structure Hosts and in MgF$_2$,* J. Chem. Phys. **86** (1987), 3455.

22. C. Reber, H. Güdel, G. Meyer, T. Schleid, and C. Daul, *Optical Spectroscopic and Structural Properties of V^{3+}-Doped Fluoride, Chloride, and Bromide Elpasolite Lattices,* Inorg. Chem. **28** (1989), 3249.

23. P. C. Schmidt, A. Weiss, and T. P. Das, *Effect of Crystal Fields and Self-Consistency on Dipole and Quadrupole Polarizabilities of Closed-Shell Ions,* Phys. Rev. **B19** (1979), 5525.

24. W. Strek, E. Lukowiak, J. Hanuza, E. Magenski, R. Cywinski, and B. Jezowka-Trzebiatowska, *Spectroscopic Properties of Cr^{3+} in Dicesium Sodium Scandium Hexachloride. The Temperature Dependence of Fluorescence,* Physica **128c** (1985), 259.

25. K. Y. Wong, N. B. Manson, and G. A. Osborne, *Magnetic Circular Dichroism of Transtion Metal Ions in Solids—II K$_2$NaGaF$_6$:Cr^{3+},* J. Phys. Chem. Solids **38** (1977), 1017.

26. R.W.G. Wyckoff, *Crystal Structures,* vol 3, Interscience, New York (1968), p 374.

5. Cs_2TiF_6

5.1 Crystallographic Data on Two Forms of Cs_2TiF_6

5.1.1 Cubic O_h^5 $(Fm3m)$, 225, Z = 4

Ion	Site	Symmetry	x^a	y	z	q	$\alpha\,(\text{Å}^3)^b$
Ti	4(a)	O_h	0	0	0	+4	0.506
Cs	8(c)	T_d	1/4	1/4	1/4	+1	2.492
F	24(e)	C_{4v}	0.195	0	0	−1	0.731

[a] X-ray data: a = 8.96 Å; the F position is not reported for Cs_2TiF_6 and is taken from Cs_2MnF_6 (Wyckoff, 1968, p 341).
[b] Schmidt et al (1979).

5.1.2 Hexagonal D_{3d}^3 $(P3m1)$, 164, Z = 1

Ion	Site	Symmetry	x^a	y	z	q	$\alpha\,(\text{Å}^3)^b$
Ti	1(a)	D_{3d}	0	0	0	+4	0.506
Cs	2(d)	C_{3v}	1/3	2/3	0.691	1	2.492
F	6(i)	C_s	0.167	−0.167	0.206	−1	0.731

[a] X-ray data, a = 6.15 Å, c = 4.96 Å; the Cs and F positions are not reported for Cs_2TiF_6 and are taken from Cs_2ZrF_6 (Wyckoff, 1968, p 350).
[b] Schmidt et al (1979).

5.2 Crystal Fields for O_h Site

5.2.1 Crystal-field components, A_{nm} (cm^{-1}/Å^n), for Ti (O_h) site of cubic form

$A_{nm}{}^a$	Monopole	Dipole	Self-induced	Total
A_{40}	25,400	32,148	−14,102	43,445

[a] $A_{44} = \sqrt{5/14}\ A_{40}$
$S^{(0)} = 18{,}682\ cm^{-1}/\text{Å}^2$
$S^{(2)} = 17{,}743\ cm^{-1}/\text{Å}^4$
$S^{(4)} = 3148.1\ cm^{-1}/\text{Å}^8$

5.2.2   **Theoretical crystal-field parameters, B_{nm} (cm^{-1}), for A_{nm} of Ti (O_h) site for quadruply ionized transition-metal ions with electronic configuration nd^N**

(a) For monopole A_{nm}

X^{4+}	N	$B_{40}{}^a$
V	1	32,969
Cr	2	20,770
Mn	3	19,713
Fe	4	12,586
Co	5	10,366
Ni	6	8,366
Cu	7	6,604
Zn	8	5,067
Ga	9	3,292

(b) For total A_{nm}

X^{4+}	N	$B_{40}{}^a$
V	1	56,392
Cr	2	35,525
Mn	3	33,718
Fe	4	21,527
Co	5	17,730
Ni	6	14,259
Cu	7	11,296
Zn	8	8,667
Ga	9	5,631

$^aB_{44} = \sqrt{5/14}\ B_{40}$

5.3   Crystal Fields for D_{3d} Site

5.3.1   **Crystal-field components, A_{nm} (cm^{-1}/Å^n), for Ti (D_{3d}) site in hexagonal form**

A_{nm}	Monopole	Dipole	Self-induced	Total
A_{20}	−5,629	−3291	1172	−7,748
A_{40}	−5,090	−3546	1986	−6,651
A_{43}	−10,051	−6316	3059	−13,308

$S^{(0)} = 8919.9\ cm^{-1}/Å^2$ $S^{(2)} = 5251.8\ cm^{-1}/Å^4$ $S^{(4)} = 460.86\ cm^{-1}/Å^8$

5.3.2   **Theoretical crystal-field parameters, B_{nm} (cm^{-1}), for A_{nm} of Ti (D_{3d}) site for quadruply ionized transition-metal ions with electronic configuration nd^N**

		(a) For monopole A_{nm}			(b) For total A_{nm}		
X^{4+}	N	B_{20}	B_{40}	B_{43}	B_{20}	B_{40}	B_{43}
V	1	−3500	−6607	13,046	−4817	−8633	17,274
Cr	2	−2911	−4162	8,219	−4007	−5439	10,882
Mn	3	−2764	−3950	7,801	−3805	−5162	10,328
Fe	4	−2228	−2522	4,980	−3067	−3296	6,594
Co	5	−2054	−2077	4,102	−2827	−2714	5,431
Ni	6	−1860	−1671	3,299	−2560	−2183	4,368
Cu	7	−1679	−1323	2,613	−2311	−1729	3,460
Zn	8	−1524	−1016	2,005	−209	−1327	2,655
Ga	9	−1290	−660	1,303	−1775	−862	1,725

5.3.3 Experimental values (cm^{-1}) of parameters for D_{3d} site of Cs_2TiF_6

Ion	nd^N	F^2	F^4	ζ	B_{20}	B_{40}	B_{43}	Ref
Mn	$3d^3$	58,058	46,557	350	−1128	−28,584	−35,571	3

5.4 Bibliography and References

1. M. J. Blandamer, J. Burgess, S. J. Hamsphere, R. D. Peacock, J. H. Rodgers, and H. D. B. Jenkins, *Thermochemistry of Hexafluoro-Anions of M^{IV} (M = Si, Ti, Mn or Re) and Lattice Energy Calculations for their Salts*, J. Chem. Soc. Dalton Trans. (1981), 726.

2. C. Campochiaro, D. S. McClure, P. Rabinowitz, and S. Dougal, *Two-Photon Spectroscopy of the $^4A_{2g} \rightarrow {}^4T_{2g}$ Transition in Mn^{4+} Impurity Ions in Crystals*, in *Vibronic Processes in Inorganic Chemistry*, C. D. Flint, ed., Kluwer Academic Publishers, Boston (1989), p 255.

3. Z. Hasan and N. B. Manson, *Jahn-Teller Effect in the $^4T_{2g}$ State of Mn^{4+} in Trigonal Cs_2TiF_6*, J. Phys. **C13** (1980), 2325.

4. N. B. Manson, G. A. Shah, B. Howes, and C. D. Flint, $^4A_g \leftrightarrow {}^2E_g$ *Transition of Mn^{4+} in $CX_2TiF_6:MnF_6^{2-}$*, Mol. Phys. **34** (1977), 1157.

5. P. C. Schmidt, A. Weiss, and T. P. Das, *Effect of Crystal Fields and Self-Consistency on Dipole and Quadrupole Polarizabilities of Closed-Shell Ions*, Phys. Rev. **B19** (1979), 5525.

6. R.W.G. Wyckoff, *Crystal Structures*, vol 3, Interscience, New York (1968).

6. NH₄Al(SO₄)₂

6.1 Crystallographic Data on $NH_4Al(SO_4)_2$

Trigonal D_3^2 ($P321$) (hexagonal setting), 150, $Z = 1$

Ion	Site	Symmetry	x^a	y	z	q	a $(Å^3)^b$
NH_4	$1(a)$	D_3	0	0	0	1	2.684
Al	$1(b)$	D_3	0	0	1/2	3	0.0530
S	$2(d)$	C_3	1/3	2/3	0.222	6	4.893
O_1	$2(d)$	C_3	1/3	2/3	0.016	−2	1.349
O_2	$6(g)$	C_2	0.328	0.344	0.317	−2	1.349

[a] X-ray data: $a = 4.724$ Å, $c = 8.225$ Å; the positions for S, O_1, and O_2 in $NH_4Al(SO_4)_2$ are not given. Those listed above are for the same ions in $KAl(SO_4)_2$ (Wyckoff, 1968).

[b] Schmidt et al (1979).

6.2 Crystal-Field Components, A_{nm} (cm⁻¹/Åⁿ), for Al (D_3) Site

Anm	Monopole	Self-induced	Dipole	Total
A_{20}	12,000	−1,983	−2,465	7,551
A_{33}	i10,394	−i1,872	i9,824	−i1,302
A_{40}	−9,527	3,347	7,693	1,512
A_{43}	−1,443	−594	11,572	9,535
A_{53}	i5,661	−i2,297	−i5,803	−i2,439

$i = \sqrt{-1}$

6.3 Experimental Parameters (cm⁻¹)

Ion	$F^{(2)}$	$F^{(4)}$	ζ	$B_{40}{}^a$	Reference
Cr^{3+}	51,818	31,019	186	−25,704	4
Ni^{3+}	54,863	32,603	600	−14,000	3

[a] Cubic approximation $B_{43} = \sqrt{10/7} \ B_{40}$

6.4 Bibliography and References

1. L. Cianchi, F. Del Giallo, P. Moretti, F. Pieralli, A. Cuccoli, M. Mancini, and G. Spina, *Mössbauer Effect Study of Relaxation Mechanism in* ^{57}*Fe-Doped Ammonium Alum*, Hyperfine Interactions **29** (1986), 1365.

2. F. Köksal, *Investigation of the H¹ Spin-Lattice Relaxation Times of Some Ammonium Compounds*, Z. Naturforsch. **34a** (1979), 239.

3. S. V. Lakshman, and K. Janardhanam, *Optical Absorption Spectrum of Ni^{2+} Doped in Ammonium Zinc Sulfate,* Pramana **11** (1978), 697.

4. S.V.J. Lakshman, B. C. Venkata Reddy, and J. Lakshmana Rao, *Crystal Field, Spin Orbit and Excitation Interactions in the Spectrum of Chromium Doped Ammonium Aluminium Sulphate,* Physica **98B** (1979), 65.

5. J. W. Mullin and M. Kitamura, *The Crystallization and Co-Crystallization of Ammonium and Potassium Aluminium Sulphates from Aqueous Solution,* J. Cryst. Growth **71** (1985), 118.

6. P. C. Schmidt, A. Weiss, and T. P. Das, *Effect of Crystal Fields and Self-Consistency on Dipole and Quadrupole Polarizabilities of Closed-Shell Ions,* Phys. Rev. **B19** (1979), 5525.

7. R.W.G. Wyckoff, *Crystal Structures,* vol 3, Interscience, New York (1968), p 166.

8. Zhang Yu-Xin, Du Mau-Lu, and Li Yong-Tai, *A Theoretical Approach of Pressure Dependence of Zero-Field Splitting of Cr^{3+} Ion in Ammonium Aluminum Alum,* Solid State Commun. **69** (1989), 945.

7. MgF$_2$

7.1 Crystallographic Data on MgF$_2$

7.1.1 Tetragonal D_{4h}^{14} ($P4_2/mnm$), 136, Z = 2

Ion	Site	Symmetry	x	y	z	q	α (Å^3)^a
Mg	2(a)	D_{2h}	0	0	0	2	0.0809
F	4(f)	C_{2v}	x	x	0	−1	0.731

^{a}Schmidt et al (1979).

7.1.2 X-ray data on MgF$_2$

a	b	x	Ref
4.623	3.052	0.303	29
4.6213	3.0519	0.303	30
4.628	3.045	0.3032	2

7.2 Crystal-Field Components, A_{nm} (cm^{-1}/Å^n), for Mg (D_{2h}) Site

7.2.1 X-ray data of Wyckoff (1968)

A_{nm}	Monopole	Self-induced	Dipole	Total
A_{20}	−576	−593	3745	2576
A_{22}	2,447	−328	−1807	313
A_{40}	−3,020	660	−382	−2742
A_{42}	−10,015	3965	−212	−6262
A_{44}	4,458	−2057	−513	1887

7.2.2 X-ray data of Yuen et al (1974)

A_{nm}	Monopole	Self-induced	Dipole	Total
A_{20}	−567	−593	3733	2573
A_{22}	2,463	−331	−1801	331
A_{40}	−3,014	658	−380	−2736
A_{42}	−10,026	3972	−211	−6264
A_{44}	4,464	−2062	−512	1890

7.2.3 X-ray data of Baur et al (1982)

A_{nm}	Monopole	Self-induced	Dipole	Total
A_{20}	−581	−615	3983	2787
A_{22}	2,262	−296	−1913	53
A_{40}	−3,128	703	−408	−2833
A_{42}	−10,038	3979	−238	−6297
A_{44}	4,439	−2044	−540	1856

7.3 Experimental Parameters (cm^{-1})

Ion	nd^N	$F^{(2)}$	$F^{(4)}$	α	$B_{40}{}^a$	Reference
V^{2+}	$3d^3$	53,362	28,065	79	24,150	24
V^{2+}	$3d^3$	49,508	34,871	—	21,735	11
Cr^{3+}	$3d^3$	—	—	—	31,374	17
Co^{2+}	$3d^7$	75,670	53,298	—	16,590	9
Ni^{2+}	$3d^8$	77,560	51,408	—	13,965	10

^{a}Cubic approximation $B_{44} = \sqrt{5/14}\, B_{40}$

7.4 Bibliography and References

1. Yu. M. Aleksandrov, V. N. Makhov, P. A. Rodnyi, T. I. Syreischikova, and M. N. Yakimenko, *Luminescence of Irradiated MgF$_2$ Crystals on Synchrotron Excitation*, Opt. Spectrosc. **61** (1986), 557.

2. W. H. Baur, S. Guggenheim, and Jiunn-Chorng Lin, *Rutile Type Compounds. VI. Refinement of VF$_2$ and Computer Simulation of V:MgF$_2$*, Acta Crystallogr. **B38** (1982), 351.

3. L. H. Brixner and G. Blasse, *X-ray Excited Emission of MgF$_2$-Gd^{3+}*, Mater. Res. Bull. **24** (1989), 441.

4. J. Ferguson, H. J. Guggenheim, L. F. Johnson, and H. Kamimura, *Magnetic Dipole Characteristics of the $^3A_{2g} \leftrightarrow {}^3T_{2g}$ Transition in Octahedral Nickel (II) Compounds*, J. Chem. Phys. **38** (1963), 2579.

5. L. F. Johnson, R. E. Dietz, and H. J. Guggenheim, *Spontaneous and Stimulated Emission from Co^{2+} Ions in MgF$_2$ and ZnF$_2$*, Appl. Phys. Lett. **5** (1964), 21.

6. L. F. Johnson, R. E. Dietz, and H. J. Guggenheim, *Optical Maser Oscillation from Ni^{2+} in MgF$_2$ Involving Simultaneous Emission of Phonons*, Phys. Rev. Lett. **11** (1963), 318.

7. L. F. Johnson and H. J. Guggenheim, *Phonon-Terminated Coherent Emission from V^{2+} Ions in MgF$_2$*, J. Appl. Phys. **38** (1967), 4837.

8. L. F. Johnson, H. J. Guggenheim, and R. A. Thomas, *Phonon-Terminated Optical Masers*, Phys. Rev. **149** (1966), 179.

9. H. Manna and R. Moncorge, *Excited-State Absorption of Co^{2+} in MgF$_2$ and KZnF$_3$*, Opt. Quant. Electron. **22** (1990), S219.

10. K. Moncorge and T. Benyattou, *Excited-State Absorption of Ni^{2+} in MgF$_2$ and MgO*, Phys. Rev. **B37** (1988), 9186.

11. K. Moncorge and T. Benyattou, *Excited State Absorption and Laser Parameters of V^{2+} in MgF$_2$ and KMgF$_3$*, Phys. Rev. **B37** (1988), 9177.

12. R. Moncorge and T. Benyattou, *Exited-State Absorption Measurements in the Ni^{2+} Doped MgO and MgF$_2$ Vibronic Laser Systems*, J. Lumin. **40,41** (1988), 105.

13. P. F. Moulton, *Tunable Solid-State Lasers Targeted for a Variety of Applications*, Laser Focus **23** (1987), 56.

14. P. F. Moulton, *An Investigation of the Co:MgF$_2$ Laser System*, IEEE J. Quant. Electron. **QE-21** (1985), 1582.

15. P. F. Moulton and A. Mooradian, *Continuously Tunable CW Ni:MgF$_2$ Laser*, 1979 IEEE/OSA Conference on Laser Engineering and Applications, p 87D.

16. S. A. Payne, L. L. Chase, and W. F. Krupke, *Optical Properties of Cr^{3+} in Fluorite-Structure Hosts and in MgF$_2$*, J. Chem. Phys. **86** (1987), 3455.

17. S. A. Payne, L. L. Chase, and G. D. Wilke, *Excited State Absorption Spectra of V^{2+} in KMgF$_3$ and MgF$_2$*, Phys. Rev. **B37** (1988), 998.

18. Yuanwu Qiu and Ji-kang Zhu, *An X$_\alpha$ Study of the Laser Crystal MgF$_2$:V^{2+}*, Z. Phys. **B75** (1989), 447.

19. J. L. Rao, R. M. Krishma, and S.V.J. Lakshman, *Nearest Neighbor Point Ion Approximation Calculations of Energy Levels for Mn^{2+} Ions Doped in MgF$_2$ Single Crystals*, Phys. Status Solidi **(b)143** (1987), K99.

20. S. Remme, G. Lehmann, R. Recker, and F. Wallrafen, *EPR and Luminescence of Mn^{2+} in MgF$_2$ Single Crystals*, Solid State Commun. **56** (1985), 73.

21. P. C. Schmidt, A. Weiss, and T. P. Das, *Effect of Crystal Fields and Self-Consistency on Dipole and Quadrupole Polarizabilities of Closed-Shell Ions*, Phys. Rev. **B19** (1979), 5525.

22. R. R. Sharma and S. Sundaram, *Transition Metal Ions in Crystals: A Refined Treatment and Deduction of Coulomb and Exchange Interaction Constants*, Solid State Commun. **33** (1979), 381.

23. W. A. Sibley, S. I. Yun, and L. N. Feuerhelin, *Radiation Defect and 3d Impurity Absorption in MgF$_2$ and KMgF$_3$ Crystals*, J. Phys. Paris **34** (1973), C9-503.

24. M. D. Sturge, F. R. Merritt, L. F. Johnson, H. J. Guggenheim, and J. P. Van der Ziel, *Optical and Microwave Studies of Divalent Vanadium in Octahedral Fluoride Coordination*, J. Chem. Phys. **54** (1971), 405.

25. R. J. Tonucci, S. M. Jacobsen, and W. M. Yen, *Dynamics of Excited States in Ni^2:MgF$_2$*, J. Lumin. **46** (1990), 155.

26. R. J. Tonucci, S. M. Jacobsen, and W. M. Yen, *Observation of the Spin-Orbit Components of the $^3B_{2g}$ ($^3A_{2g}$) Ground State in the System Ni^{2+}:MgF$_2$ by Fluorescence Line Narrowing*, Chem. Phys. Lett. **173** (1990), 456.

27. G. Vidal-Valat, J.-P. Vidal, and C. M. E. Zeyen, *On the Atomic Vibrations in the Magnesium Difluoride Crystal*, Acta Crystallogr. **B36** (1980), 2857.

28. G. Vidal-Valat, J.-P. Vidal, C. M. E. Zeyen, and K. Kurki-Suonio, *Neutron Diffraction Study of Magnesium Fluoride Single Crystals,* Acta Crystallogr. **B35** (1979), 1584.

29. N. W. Winter and R. M. Pitzer, *Configuration Interaction Calculation of the Electronic Spectra of MgF$_2$:V^{+2},* J. Chem. Phys. **89** (1988), 446.

30. R.W.G. Wyckoff, *Crystal Structures,* vol 1, Interscience, New York (1968), p 251.

31. P. S. Yuen, R. M. Murfitt, and R. L. Collin, *Interionic Forces and Ionic Polarization in Alkaline Earth Halide Crystals,* J. Chem. Phys. **61** (1974), 2383.

32. S. I. Yun, L. A. Kappers, and W. A. Sibley, *Enhancement of Impurity Ion Absorption due to Radiation-Produced Defects,* Phys. Rev. **B5** (1973), 773.

33. B. Zhang, J.-K. Zhy, and S.-H. Liu, *Crystal Field Energy Levels and Optical Absorption Intensities of Ni^{2+}:MgF$_2$,* NBS Meeting on Basic Properties of Optical Materials (7–9 May 1985).

8. MnF$_2$

8.1 Crystallographic Data on MnF$_2$

8.1.1 Tetragonal D_{4h}^{14} ($P4_2/mnm$), 136, Z = 2

Ion	Site	Symmetry	x	y	z	q	α (Å^3)^b
Mn	2(a)	D_{2h}	0	0	0	+2	0.122
F	4(f)	C_{2v}	x	x	0	−1	0.731

^{a}Schmidt et al (1979).

8.1.2 X-ray data on MnF$_2$

a	c	x	Ref
4.8734	3.3099	0.305	1,19
4.8734	3.3099	0.310	18

8.2 Crystal Fields for Mn (D_{2h}) Site

8.2.1 Crystal-field components, A_{nm} (cm^{-1}/Å^n), for Mn (D_{2h}) site (Baur, 1958; Wyckoff, 1968)

A_{nm}	Monopole	Self-induced	Dipole	Total
A_{20}	901	−459	1816	2258
A_{22}	2638	−301	−847	1490
A_{40}	−1670	267	−125	−1529
A_{42}	−7263	2376	−62	−4948
A_{44}	3218	−1240	−223	1755

8.2.2 Crystal-field components, A_{nm} (cm^{-1}/Å^n), for Mn (D_{2h}) site (Stout and Reed, 1954)

A_{nm}	Monopole	Self-induced	Dipole	Total
A_{20}	2878	−711	2621	4788
A_{22}	1761	−160	−1105	496
A_{40}	−1781	364	−113	−1530
A_{42}	−7338	2419	−110	−5029
A_{44}	2989	−1128	−287	1574

8.3   Experimental Parameters (cm^{-1})

Ion	$F^{(2)}$	$F^{(4)}$	α	ζ	$B_{40}{}^a$	Reference
Mn^{2+}	62,230	39,690	76	320	15,750	10
Mn^{2+}	61,355	39,791	76	320	15,792	10
Mn^{2+}	58,849	46,746	—	—	17,220	4
Mn^{2+}	67,830	41,215	66	—	16,569	16
Mn^{2+}	67,550	41,240	65	337	17,300	20[b]
Mn^{2+}	69,510	41,328	—	—	16,380	17
Co^{2+}	68,159	42,424	—	—	17,220	2

^{a}*Cubic approximation* $B_{44} = \sqrt{5/14}\, B_{40}$

^{b}*Fit with full* D_{2h} *symmetry:* $B_{20} = -1480$, $B_{22} = 4750$, $B_{42} = -1650$, $B_{44} = -10,260$ *(cm^{-1}).*

8.4   Bibliography and References

1. W. H. Baur, *Uber die Verfeinerung der Kristallstrukturbestimmung einieger Vertreter des Rutiltyps: II.—Die Difluoride von Mn, Fe, Co, Ni und Zn*, Acta Crystallogr. **11** (1958), 488.

2. L. F. Blunt, *Optical Absorption of Cobalt in Manganese Fluoride*, J. Chem. Phys. **44** (1966), 2317.

3. D. Curie, C. Barthou, and B. Canny, *Covalent Bonding of Mn^{2+} Ions in Octahedral and Tetrahedral Coordination*, J. Chem. Phys. **61** (1974), 3048.

4. D. M. Finlayson, I. S. Robertson, T. Smith, and R.W.H. Stevenson, *The Optical Absorption of Manganese Zinc Fluoride Single Crystals Near the Néel Temperature*, Proc. Phys. Soc. **76** (1960), 355.

5. H. J. Hrostowski and R. H. Kaiser, *Absorption Spectra of MnF$_2$ and KMnF$_3$*, Bull. Am. Phys. Soc. **4** (1959), 167.

6. L. F. Johnson, R. E. Dietz, and H. J. Guggenheim, *Exchange Splitting of the Ground State of Ni^{2+} Ions in Antiferromagnetic MnF$_2$, KMnF$_3$, and RbMnF$_3$*, Phys. Rev. Lett. **17** (1966), 13.

7. L. F. Johnson, R. E. Dietz, and H. J. Guggenheim, *Optical Maser Oscillation from Ni^{2+}in MgF$_2$ Involving Simultaneous Emission of Phonons*, Phys. Rev. Lett. **11** (1963), 318.

8. L. F. Johnson, H. J. Guggenheim, and R. A. Thomas, *Phonon-Terminated Optical Masers*, Phys. Rev. **149** (1966), 179.

9. N. A. Kulagin and D. T. Sviridov, *Charge of the Effective Occupation Numbers of 3d-Shells of Mn^{2+} Ions in Manganese Halide Crystals*, Phys. Status Solidi **(b)112** (1982), K61.

10. W. Low and G. Rosengarten, *The Optical Spectrum and Ground-State Splitting of Mn^{2+} and Fe^{3+} Ions in the Crystal Field of Cubic Symmetry,* J. Mol. Spectrosc. **12** (1964), 319.

11. D. S. McClure, R. Meltzer, S. A. Reed, P. Russell, and J. W. Stout, *Electronic Transitions with Spin Change in Several Antiferromagnetic Crystals,* in *Optical Properties of Ions in Crystals,* Interscience, New York (1967), p 257.

12. B. Ng and D. J. Newman, *Ab Initio Calculations of Ligand Field Correlations Effects in Mn^{2+}-F$^-$,* Chem. Phys. Lett. **130** (1986), 410.

13. B. Ng and D. J. Newman, *A Linear Model of Crystal Field Correlation Effects in Mn^{2+},* J. Chem. Phys. **84** (1986), 3291.

14. J. Quazza and J. Labbe, *Theoretical Models for the Raman Scattering by Localized and Magnon Modes in Mixed Transition-Metal Fluorides,* Phys. Rev. **B34** (1986), 8449.

15. P. C. Schmidt, A. Weiss, and T. P. Das, *Effect of Crystal Fields and Self-Consistency on Dipole and Quadrupole Polarizabilities of Closed-Shell Ions,* Phys. Rev. **B19** (1979), 5525.

16. R. Stevenson, *Absorption Spectra of MnF$_2$, KMnF$_2$, RbMnF$_2$, and CsMnF$_2$,* Can. J. Phys. **43** (1965), 1732.

17. J. W. Stout, *Absorption Spectrum of Manganese Fluoride,* J. Chem. Phys. **31** (1959), 709.

18. J. W. Stout and S. A. Reed, *The Crystal Structure of MnF$_2$, FeF$_2$, CoF$_2$, NiF$_2$, and ZnF$_2$,* J. Am. Chem. Soc. **76** (1954), 5279.

19. R.W.G. Wyckoff, *Crystal Structures,* vol 1, Interscience, New York (1968), p 251.

20. W.-L. Yu and M.-G. Zhao, *Determination of the MnF$_2$ and ZnF$_2$ Crystal Structure from the EPR and Optical Measurement of Mn^{2+},* J. Phys. **C18** (1985), L1087.

9. ZnF$_2$

9.1 Crystallographic Data on ZnF$_2$

Tetragonal D_{4h}^{14} ($P4_2/mnm$), 136, $Z = 2$

Ion	Site	Symmetry	x	y	z	q	a (Å^3)^a
Zn	2(a)	D_{2h}	0	0	0	+2	0.676
F	4(f)	C_{2v}	x	x	0	−1	0.731

^{a}Schmidt et al (1979).

X-ray data

a	c	x	Ref
4.7034	3.1335	0.303	1,14
4.7034	3.1335	0.307	12

9.2 Crystal Fields for Zn Site (D_{2h})

9.2.1 X-ray data of Baur (1958), Wyckoff (1968)

A_{nm}	Monopole	Self-induced	Dipole	Total
A_{20}	−305	−593	2855	1957
A_{22}	2659	−346	−1379	934
A_{40}	−2491	430	−268	−2329
A_{42}	−9011	3350	−135	−5796
A_{44}	4046	1786	−383	1876

$S^{(0)} = 8214.0\ cm^{-1}/Å^2$

$S^{(2)} = 5417.7\ cm^{-1}/Å^4$

$S^{(4)} = 508.50\ cm^{-1}/Å^8$

9.2.2 X-ray data of Stout and Reed (1954)

A_{nm}	Monopole	Self-induced	Dipole	Total
A_{20}	1476	−850	3761	4387
A_{22}	1832	−195	−1676	−38
A_{40}	−2638	557	−269	−2350
A_{42}	−9102	3405	−208	−5905
A_{44}	−3817	−1660	−460	1698

9.3   Experimental Parameters (cm^{-1})

Ion	$F^{(2)}$	$F^{(4)}$	α	ζ	B_{20}	B_{22}	B_{40}	B_{42}	B_{44}	Ref
Mn^{2+}	67,893	41,240	65	0	0	0	19,320^a	0	—	4
Mn^{2+}	57,550	41,240	65	337	−920	3730	21,180	−1390	−12,660	15
Mn^{2+}	—	—	—	—	0	0	19,320^a	0	—	9
Ni^{2+}	64,400	34,776	—	—	0	0	15,750^a	0	—	8

$^a B_{44} = \sqrt{5/14}\ B_{40}$

9.4   Bibliography and References

1. W. H. Baur, *Uber die Verfeinerung der Kristallstrukturbestimmung einieger Vertreter des Rutiltyps: II. —Die Difluoride von Mn, Fe, Co, Ni, und Zn,* Acta Crystallogr. **11** (1958), 488.

2. C. Binoit and J. Giordano, *Dynamical Properties of Crystals of MgF$_2$, ZnF$_2$ and FeF$_2$: II. Lattice Dynamics and Infrared Spectra,* J. Phys. **C21** (1988), 5209. (Part I is Giordano and Benoit, J. Phys. **C21** (1988), 2749.)

3. Jia-Jun Chen and Min-Guang Zhao, *High-Order Perturbation Formula of the Spin-Hamiltonian Parameters for the Ground State $^3A_2$ (F) of d^8 Ions in Rhombic Symmetry,* Phys. Status Solidi **(b)143** (1987), 647.

4. D. Curie, C. Barthou, and B. Canny, *Covalent Bonding of Mn^{2+} Ions in Octahedral and Tetrahedral Coordination,* J. Chem. Phys. **61** (1974), 3048.

5. D. M. Finlayson, I. S. Robertson, T. Smith, and R.W.H. Stevenson, *The Optical Absorption in Manganese Zinc Fluoride Single Crystals Near the Néel Temperature,* Proc. Phys. Soc. **76** (1960), 355.

6. L. F. Johnson, R. E. Dietz, and H. J. Guggenheim, *Spontaneous and Stimulated Emission from Co^{2+} Ions in MgF$_2$ and ZnF$_2$,* Appl. Phys. Lett. **5** (1964), 21.

7. L. F. Johnson, H. J. Guggenheim, and R. A. Thomas, *Phonon-Terminated Optical Masers,* Phys. Rev. **149** (1966), 179.

8. A. D. Liehr, *The Three Electron (or Hole) Cubic Ligand Field Spectrum,* J. Phys. Chem. **67** (1963), 1314.

9. D. T. Palumbo and J. J. Brown, *Electronic States of Mn^{2+} Activated Phosphors: II.—Orange-to-Red Emitting Phosphors,* J. Electrochem. Soc. **118** (1971), 1159.

10. M. L. Reynolds and G.F.J. Garlick, *The Infrared Emission of Nickel Ion Impurity Centers in Various Solids,* Infrared Phys. **7** (1967), 151.

11. P. C. Schmidt, A. Weiss, and T. P. Das, *Effect of Crystal Fields and Self-Consistency on Dipole and Quadrupole Polarizabilities of Closed-Shell Ions,* Phys. Rev. **B19** (1979), 5525.

12. J. W. Stout and S. A. Reed, *The Crystal Structure of MnF_2, FeF_2, CoF_2, NiF_2, and ZnF_2*, J. Am. Chem. Soc. **76** (1954), 5279.

13. M. Tinkham, *Paramagnetic Resonance in Dilute Iron Group Fluorides: I.—Fluorine Hyperfine Structure*, Proc. Roy. Soc. **A236** (1956), 535.

14. R.W.G. Wyckoff, *Crystal Structures*, vol 1, Interscience, New York (1968), p 251.

15. W.-L. Yu and M.-G. Zhao, *Determination of the MnF_2 and ZnF_2 Crystal Structure from the EPR and Optical Measurements of Mn^{2+}*, J. Phys. **C18** (1985), L1087.

10. MgO

10.1 Crystallographic Data on MgO

Cubic O_h^5 ($Fm3m$), 225, $Z = 4$

Ion	Site	Symmetry	x^a	y	z	q	α (Å³)^b
Mg	4(a)	O_h	0	0	0	+2	0.0809
O	4(b)	O_h	1/2	1/2	1/2	−2	1.349

^{a}X-ray data, $a = 4.2112$ (Wyckoff, 1964).
^{b}Schmidt et al (1979).

10.2 Crystal Fields for Mg (O_h) Site

10.2.1 Crystal-field components, A_{nm} (cm^{-1}/Å^n), for Mg (O_h) site

A_{nm}^{a}	Monopole	Self-induced	Total
A_{40}	20,084	−5,812	14,271

$^a A_{44} = \sqrt{5/14}\, A_{40}$
$S^{(0)} = 11,851\ cm^{-1}/Å^2$
$S^{(2)} = 7523.7\ cm^{-1}/Å^4$
$S^{(4)} = 621.35\ cm^{-1}/Å^8$

10.2.2 Theoretical crystal-field parameters, B_{nm} (cm^{-1}), for A_{nm} of Mg (O_h) site for transition-metal ions with electronic configuration nd^N

X^{2+}	N	For monopole A_{nm}^{a} B_{40}^{b}	For total A_{nm}^{a} B_{40}^{b}
Sc	1	81,400	57,840
Ti	2	50,310	35,749
V	3	34,504	24,518
Cr	4	24,784	17,610
Mn	5	18,021	12,805
Fe	6	14,533	10,326
Co	7	11,524	8,188
Ni	8	9,192	6,532
Cu	9	7,387	5,249

$^a B_{40} = <r^4>_{H-F}\, A_{40}$
$^b B_{44} = \sqrt{5/14}\, B_{40}$

10.3   Experimental Parameters (cm^{-1})

Ion	nd^N	$F^{(2)}$	$F^{(4)}$	α	ζ	B_{40}	Ref
V^{2+}	$3d^3$	42,429	30,239	60	—	30,429	59
V^{2+}	$3d^3$	47,625	25,455	79	—	29,400	55
V^{2+}	$3d^3$	44,275	31,185	—	—	27,720	56
Cr^{3+}	$3d^3$	50,906	37,825	70	—	33,579	59
Cr^{3+}	$3d^3$	52,780	36,792	70	270	33,390	7
Cr^{3+}	$3d^3$	54,600	40,950	—	135	34,020	28
Cr^{3+}	$3d^3$	54,250	40,320	—	210	34,860	36
Cr^{2+}	$3d^4$	—	—	—	—	14,007	12
Mn^{2+}	$3d^5$	60,984	40,446	—	—	20,559	11
Mn^{2+}	$3d^5$	67,550	41,240	65	320	20,580	23
Mn^{2+}	$3d^5$	62,230	39.690	76	320	15,750	22
Co^{2+}	$3d^7$	68,572	49,342	—	419	19,509	61
Co^{2+}	$3d^7$	65,450	47,250	—	500	18,795	38
Co^{2+}	$3d^7$	67.060	46,620	—	475	19,530	50
Ni^{2+}	$3d^8$	—	—	—	—	18,060	30
Ni^{2+}	$3d^8$	71,022	49,342	—	630	17,115	46
Ni^{2+}	$3d^8$	66,343	44,447	—	645	17,451	42

10.4   Bibliography and References

1. B. R. Anderson, L. J. Challis, J.H.M. Stoelinga, and P. Wyder, *Far Infrared Studies of Cr^{2+} in MgO*, J. Phys. **C7** (1974), 2234.

2. K. W. Blazey, *Optical Absorption of MgO:Fe*, J. Phys. Chem. Solids **38** (1977), 671.

3. A. Boyrivent, E. Duval, M. Montagna, G. Viliani, and O. Pilla, *New Experimental Results for the Interpretation of the $^4A_{2g} \rightarrow {}^4T_{2g}$ Spectra of Cr^{3+} and V^{2+} in MgO*, J. Phys. **C12** (1979), L803.

4. L. L. Chase, *Electron Spin Resonance of the Excited $2E(3d^3)$ Level of Cr^{3+} and V^{2+} in MgO*, Phys. Rev. **168** (1968), 341.

5. J. J. Davies and J. E. Wertz, *Trivalent Titanium in Tetragonal Symmetry in MgO and CaO*, J. Magn. Reson. **1** (1969), 500.

6. C.G.C.M. DeKort, *Millimetre and Submillimetre Wave Spectroscopy of Solids*, doctoral thesis (1979), De Katholieke Universiteit Te Nijmegen, chapters III, IV.

7. W. M. Fairbank, Jr., and G. K. Klauminzer, *Tetragonal-Field Splittings of Levels in MgO:Cr^{3+}*, Phys. Rev. **B7** (1973), 500.

8. E. R. Feher, *Effect of Uniaxial Stresses on the Paramagnetic Spectra of Mn^{3+} and Fe^{3+} in MgO*, Phys. Rev. **136** (1964), A145.

9. J. Ferguson, K. Knox, and D. L. Wood, *Effect of Next Nearest Neighbor Ions on the Crystal Field Splitting of Transition Metal Ions in Crystals*, J. Chem. Phys. **35** (1961), 2236.

10. J. R. Fletcher and K.W.H. Stevens, *The Jahn-Teller Effect of Octahedrally Co-ordinated $3d^4$ Ions*, J. Phys. **C2** (1969), 444.

11. J. R. Gabriel, D. F. Johnston, and M.J.D. Powell, *A Calculation of the Ground State Splitting for Mn^{2+} Ions in a Cubic field*, Proc. Roy. Soc. (1961), 503.

12. C. Greskovich and V. S. Stubican, *Divalent Chromium in Magnesium-Chromium Spinels*, J. Phys. Chem. Solids **27** (1966), 1379.

13. F. Hasan, P. J. King, D. J. Monk, D. T. Murphy, V. W. Rampton, and P. C. Wiscombe, *Phonon Spectroscopy of Mn^{3+} in MgO*, in *Phonon Scattering in Condensed Matter*, H. J. Marris, ed., Third International Conference on Phonon Scattering in Condensed Matter, Brown University (1979), p 85.

14. F. S. Ham, *Acoustic Paramagnetic Resonance Spectrum of Cr^{2+} in MgO*, Phys. Rev. **B4** (1971), 3854.

15. F. S. Ham, W. M. Schwartz, and M.C.M. O'Brien, *Jahn-Teller Effects in the Far-Infrared, EPR, and Mossbauer Spectra of $MgO:Fe^{2+}$*, Phys. Rev. **185** (1969), 548.

16. A. Hjortrsberg, J. T. Vallin, and F. S. Ham, *Jahn-Teller Effects in the Near-Infrared Absorption Spectrum of $MgO:Fe^{2+}$*, Phys. Rev. **B37** (1988), 3196.

17. U. Höchli, K. A. Müller, and P. Wysling, *Paramagnetic Resonance and Relaxation of Cu^{2+} and Ni^{3+} in MgO and CaO: The Determination of Jahn-Teller Energy Splittings*, Phys. Lett. **15** (1965), 5.

18. C. Y. Huang, R. S. Kent, and S. A. Marshall, *Temperature Dependence of the Hyperfine-Structure Splitting of Divalent Manganese in Single-Crystal Magnesium Oxide*, Phys. Rev. **B7** (1973), 552.

19. J. L. Janson and Z. A. Rachko, *Nature of Impurity-Induced UV Luminescence of MgO Crystals*, Phys. Status Solidi **(a)53** (1979), 121.

20. L. F. Johnson, R. E. Dietz, and H. J. Guggenheim, *Optical Maser Oscillation from Ni^{2+} in MgF_2 Involving Simultaneous Emission of Phonons*, Phys. Rev. Lett. **11** (1963), 318.

21. L. F. Johnson, H. J. Guggenheim, and R. A. Thomas, *Phonon-Terminated Optical Masers*, Phys. Rev. **149** (1966), 179.

22. P. Koidl and K. W. Blazey, *Optical Absorption of MgO:Mn*, J. Phys. **C9** (1976), L167.

23. X.-Y. Kuang, *Stark Levels $^4T_1(G)$ in $MgO:Mn^{2+}$*, Phys. Rev. **B37** (1988), 9719.

24. M. Kunz, H. Kretschmann, W. Assmus, and C. Klingshirn, *Absorption and Emission Spectra of Yttria-Stabilized Zirconia and Magnesium Oxide*, J. Lumin. **37** (1987), 123.

25. R. Lacroix, J. Weber, E. Duval, and A. Boyrivent, *Structure of the $^4A_{2g} \rightarrow {}^4T_{2g}$ Zero-Phonon Spectrum in MgO:Cr^{3+}. Spin-Orbit and Vibronic Coupling*, J. Phys. **C13** (1980), L781.

26. J. Lange, *Dynamic Jahn-Teller Effect for Cr^{2+} in MgO: Acoustic Paramagnetic Resonance*, Phys. Rev. **B14** (1976), 4791.

27. J. P. Larkin, G. F. Imbusch, and F. Dravneiks, *Optical Absorption in MgO:Cr^{3+}*, Phys. Rev. **B7** (1973), 495.

28. A. D. Liehr, *The Three Electron (or Hole) Cubic Ligand Field Spectrum*, J. Phys. Chem. **67** (1963), 1314.

29. W. Low, *Paramagnetic Resonance in Solids*, Solid State Phys. Suppl. **2** (1960), 104.

30. W. Low, *Paramagnetic and Optical Spectra of Divalent Nickel in Cubic Crystalline Fields*, Phys. Rev. **109** (1958), 247.

31. W. Low and M. Weger, *Paramagnetic Resonance and Optical Spectra of Divalent Iron in Cubic Fields: I.—Theory*, Phys. Rev. **118** (1960), 1119.

32. W. Low and M. Weger, *Paramagnetic Resonance and Optical Spectra of Divalent Iron in Cubic Fields: II.—Experimental Results*, Phys. Rev. **118** (1960), 1130.

33. Z. Luz, A. Raizman, and J. T. Suss, *Oxygen-17 Superfine Structure of Rh^{2+} Jahn-Teller Ions in MgO*, Solid State Commun. **21** (1977), 849.

34. W. C. Mackrodt, R. F. Stewart, J. C. Campbell, and I. H. Hillier, *The Calculated Defect Structure of ZnO*, J. Phys. Paris **41** (Suppl. to No. 7) (1980), C6-64.

35. R. M. Macfarlane, *Excited-State Absorption Cross Sections for the Cr^{3+} Ion in MgO and Spinal*, Phys. Rev. **B7** (1971), 2129.

36. R. M. Macfarlane, *Perturbation Methods in the Calculation of Zeeman Interactions and Magnetic Dipole Line Strengths for d^3 Trigonal-Crystal Spectra*, Phys. Rev. **B1** (1970), 989. Errata, Phys. Rev. **B3** (1971), 1054.

37. N. B. Manson and M. D. Sturge, *Absence of a Strong Jahn-Teller Effect in the $^4T_{2g}$ Excited State of V^{2+} in MgO*, Phys. Rev. **B22** (1980), 2861.

38. A. J. Mann and P. J. Stephens, *Magnetic Circular Dichroism of Impurities in Solids: MgO:Co*, Phys. Rev. **B9** (1974), 863.

39. F. G. Marshall and V. W. Rampton, *The Acoustic Paramagnetic Resonance Spectrum of Chromous Ions in Magnesium Oxide*, J. Phys. **C1** (1968), 594.

40. J. Meng, J. M. Vail, A. M. Stoneham, and P. Jena, *Charge-State Stability of Ni and Cu Impurities in MgO*, Phys. Rev. **B42** (1990), 1156.

41. D. Meyer, M. Regis, and Y. Farge, *Far Infrared Absorption of Fe^{2+} Ions in MgO Crystals*, Phys. Lett. **48A** (1974), 41.

42. K. Moncorge and T. Benyattou, *Excited-State Absorption of Ni^{2+} in MgF_2 and MgO*, Phys. Rev. **B37** (1988), 9186.

43. R. Moncorge and T. Benyattou, *Excited-State Absorption Measurements in the Ni^{2+} Doped MgO and MgF_2 Vibronic Laser Systems*, J. Lumin. **40,41** (1988), 105.

44. M.C.M. O'Brien, *The Jahn-Teller Coupling of $3d^6$ Ions in a Cubic Crystal*, Proc. Phys. Soc. **86** (1965), 847.

45. M. Okada, T. Kawakubo, T. Seiyama, and M. Nakagawa, *Enhancement of 3d-Electron Transitions in Neutron-Irradiated $MgO:Mn^{2+}$ Crystals*, Phys. Status Solidi **(b)144** (1987), 903.

46. R. Pappalardo, D. L. Wood, and R. C. Linares, Jr., *Optical Absorption Spectra of Ni-Doped Oxide Systems. I*, J. Chem. Phys. **35** (1961), 1460.

47. J. L. Patel and J. K. Wigmore, *Jahn-Teller Parameters of $MgO:Cr^{2+}$ Determined Using Heat Pulses*, J. Phys. **C10** (1977), 1829.

48. S. A. Payne, *Energy-Level Assignments for the ^1E and ^3T_1a States of $MgO:Ni^{2+}$*, Phys. Rev. **B41** (1990), 6109.

49. J. E. Ralph and M. G. Townsend, *Fluorescence and Absorption Spectra of Ni^{2+} in MgO*, J. Phys. **C3** (1970), 8.

50. J. E. Ralph and M. G. Townsend, *Near-Infrared Fluorescence and Absorption Spectra of Co^{2+} and Ni^{2+} in MgO*, J. Chem. Phys. **48** (1968), 149.

51. M. Régis, Y. Farge, and M. Fontana, *Etude des transitions optiques dans $MgO:Ni^{2+}$ par dichroisme circulaire magnetique*, Phys. Status Solidi **(b)57** (1973), 307.

52. P. C. Schmidt, A. Weiss, and T. P. Das, *Effect of Crystal Fields and Self-Consistency on Dipole and Quadrupole Polarizabilities of Closed-Shell Ions*, Phys. Rev. **B19** (1979), 5525.

53. A. G. Shenstone, *The Third Spectrum of Nickel (Ni III)*, J. Opt. Soc. Am. (1954), 749.

54. G. Smith, *Evidence for Optical Absorption by Fe^{2+}-Fe^{3+} Interactions in MgO:Fe*, Phys. Status Solidi **(a)61** (1980), K191.

55. M. D. Sturge, *Jahn-Teller Effect in the $^4T_{2g}$ Excited State of V^{2+} in MgO*, Phys. Rev. **140** (1965), A880.

56. M. D. Sturge, *Optical Spectrum of Divalent Vanadium in Octahedral Coordination*, Phys. Rev. **130** (1963), 639.

57. M. D. Sturge, *Strain-Induced Splitting of the ^2E State of V^{2+} in MgO*, Phys. Rev. **131** (1963), 1456.

58. M. D. Sturge, F. R. Merritt, L. F. Johnson, H. J. Guggenheim, and J. P. van der Ziel, *Optical and Microwave Studies of Divalent Vanadium in Octahedral Fluoride Coordination*, J. Chem. Phys. **54** (1971), 405.

59. D. T. Sviridov, R. K. Sviridova, N. I. Kulik, and V. B. Glasko, *Optical Spectra of the Isoelectronic Ions V^{2+}, Cr^{3+}, and Mn^{4+} in an Octahedral Coordination*, J. Appl. Spectrosc. **30** (1979), 334.

60. G. Viliani, M. Montagna, O. Pilla, A. Fontana, M. Bacci, and A. Ranfagni, *Magnetic Circular Dichroism in Split Zero-Phonon Line of MgO:V^{2+}*, J. Phys. **C11** (1978), L439.

61. V. Wagner and P. Koidl, *Evidence of Weak Jahn-Teller Effect of Co^{2+} in MgO*, J. Magn. Magn. Mater. **15–18** (1980), 33.

62. J. Y. Wong, *Far-Infrared Spectra of Iron-Doped MgO*, Phys. Rev. **168** (1968), 337.

63. R.W.G. Wyckoff, *Crystal Structures*, vol 1, Interscience, New York (1964), p 88.

64. P. Wysling, K. A. Muller, and U. Höchli, *Paramagnetic Resonance and Relaxation of Ag^{2+} and Pd^{3+} in MgO and CaO*, Helv. Phys. Acta **38** (1965), 358.

65. M. Yamaga, B. Henderson, A. Marshall, K. P. O'Donnell, and B. Cockayne, *Polarization Spectroscopy of Cr^{3+} Ions in Laser Host Crystals: I. R Lines and Sidebands*, J. Lumin. **43** (1989), 139.

66. M. Yamaga, B. Henderson, and K. P. O'Donnell, *Polarization Spectroscopy of Cr^{3+} Ions in Laser Host Crystals: II. The Broadband Transitions*, J. Lumin. **46** (1990), 397.

67. W. M. Yen, L. R. Elias, and D. L. Huber, *Utilization of Near- and Vacuum-Ultraviolet Synchrotron Radiation for the Excitation of Visible Fluorescences in Ruby and MgO:Cr^{3+}*, Phys. Rev. Lett. **24** (1970), 1011.

68. Y. Y. Yeung, *Local Distortion and Zero-Field Splittings of $3d^5$ Ions in Oxide Crystals*, J. Phys. **C21** (1988), 2453.

69. W. -C. Zheng, *Theoretical Explaination of the Spin-Lattice Coupling Coefficients G_{11} and G_{44} for d^3 (V^{2+}, Cr^{3+}) Ions in MgO Crystals*, J. Phys. **C1** (1989), 8093.

11. $Be_3Al_2(SiO_3)_6$ (Beryl, Emerald)

11.1 Crystallographic and X-Ray Data on $Be_3Al_2(SiO_3)_6$

Hexagonal D_{6h}^2 ($P6/mcc$), 192, $Z = 2$

Ion	Site	Symmetry	x^a	y	z	q	α (Å^3)^b
Al	4(c)	D_3	1/3	2/3	1/4	+3	0.0530
Be	6(f)	D_2	1/4	0	1/4	+2	0.0125
Si	12(P)	C_s	0.382	0.118	0	+4	0.0165
O_1	12(P)	C_s	0.294	0.242	0	–2	1.349
O_2	24(m)	C_1	0.499	0.143	0.138	–2	1.349

[a] *X-ray data: a = 9.206 Å, c = 9.205 Å (Wyckoff 1968).*
[b] *Schmidt et al (1979).*

11.2 Crystal Fields for Al (D_3) Site

11.2.1 Crystal-field components, A_{nm} (cm^{-1}/Å^n), for Al (D_3) site

A_{nm}	Monopole	Self-induced	Dipole	Total
A_{20}	–16,578	1289	13,630	–1,659
A_{33}	–i4,113	i2339	–i2,941	–i4,716
A_{40}	–16,436	6117	–23,937	–34,257
A_{43}	20,357	–8341	29,273	41,288
A_{53}	–i4,004	i1543	–i3,056	–i5,516

$S^{(0)} = 18{,}026$ cm^{-1}/Å^2
$S^{(2)} = 13{,}560$ cm^{-1}/Å^4
$S^{(4)} = 1{,}535.7$ cm^{-1}/Å^8
$i = \sqrt{-1}$

11.2.2 Theoretical crystal-field parameters, B_{nm} (cm^{-1}), for A_{nm} of Al (D_3) site for transition-metal ions with electronic configuration nd^N

(a) For monopole A_{nm}^{a}

X^{3+}	N	B_{20}	B_{40}	B_{43}
Ti	1	–13,193	–21,055	26,077
V	2	–11,089	–15,031	18,616
Cr	3	–9,576	–11,359	14,069
Mn	4	–8,379	–8,815	10,917
Fe	5	–7,354	–6,865	8,503
Co	6	–6,664	–5,763	7,137
Ni	7	–6,013	–4,771	5,910
Cu	8	–5,438	–3,966	4,912
Zn	9	–4,935	–3,314	4,104

[a] $B_{nm} = \langle r^n \rangle_{H-F} A_{nm}$

(b) For total A_{nm}^{a}

B_{20}	B_{40}	B_{43}
–1320	–43,883	52,890
–1110	–31,328	37,758
–958	–23,675	28,534
–838	–18,372	22,143
–736	–14,309	17,246
–667	–12,011	14,476
–602	–9,945	11,986
–544	–8,266	9,963
–494	–6,906	8,324

11.3 Experimental Parameters (cm⁻¹)

Ion	nd^N	$F^{(2)}$	$F^{(4)}$	α	ζ	B_{20}	B_{40}	B_{43}	Ref
V^{3+}	$3d^2$	45,780	31,248	—	—	—	$-24,500^a$	—	26
V^{2+}	$3d^3$	59,500	40,950	—	—	—	$-21,000^a$	—	26
Cr^{3+}	$3d^3$	55,510	37,296	70	—	$-5,453$	$-21,123$	28,211	10
Cr^{3+}	$3d^3$	58,940	37,296	—	225	$-7,657$	$-19,690$	28,359	18
Cu^{2+}	$3d^9$	—	—	—	—	—	$-18,620^a$	—	29

^{a}Cubic approximation $B_{43} = \sqrt{10/7}\, B_{40}$

11.4 Bibliography and References

1. A. A. Akhumyan, Zh. A. Arakelyan, G. V. Bukin, R. M. Martirosyan, and V. K. Ogneva, *Quantum Parametric Amplifier Utilizing Synthetic Emerald Crystals,* Sov. J. Quantum Electron. **9** (1979), 61.

2. A. A. Akhumyan, R. M. Martirosyan, and N. G. Pogosyan, *Quantum Amplification of Millimeter Waves by Synthetic Emerald Crystals,* Sov. Tech. Phys. Lett. **7** (1981), 371.

3. P. J. Beckwith and E. J. Troup, *The Optical and Infrared Absorption of V^{3+} in Beryl ($Be_3Al_2Si_6O_{18}$),* Phys. Status Solidi **(a)16** (1973), 181.

4. J. Buchert and R. R. Alfano, *Emerald—A New Gem Laser Material,* Laser Focus (September 1983), 117.

5. J. Buchert, A. Katz, and R. R. Alfano, *Laser Action in Emerald,* IEEE J. Quantum Electron. **QE-19** (1983), 1477.

6. G. Burns, E.A. Geiss, B. A. Jenkins, and M. I. Nathan, *Cr^{3+} Fluorescence in Garnets and Other Crystals,* Phys. Rev. **139** (1965), A1687.

7. A. Edgar and D. R. Hutton, *Exchange-Coupled Pairs of Fe^{3+} Ions in Beryl,* Solid State Commun. **41** (1982), 195.

8. A. Edgar and D. R. Hutton, *Exchange-Coupled Pairs of Cr^{3+} Ions in Emerald,* J. Phys. **C11** (1978), 5051.

9. E. N. Emel'yanova, S. V. Grum-Grzhimailo, O. N. Boksha, and T. M. Varina, *Artificial Beryl Containing V, Mn, Co, and Ni,* Sov. Phys. Crystallogr. **10** (1965), 46.

10. W. M. Fairbank, Jr., G. K. Klauminzer and A. L. Schawlow, *Excited-State Absorption in Ruby, Emerald, and $MgO:Cr^{3+}$,* Phys. Rev. **B11** (1975), 60.

11. W. H. Fonger and C. W. Struck, *Temperature Dependences of Cr^{3+} Radiative and Nonradiative Transitions in Ruby and Emerald,* Phys. Rev. **B11** (1975), 3251.

12. T. Furusaki, Y. Bando, K. Kodaira, and T. Matsushita, *Properties of Beryl Single Crystals Grown by a High Pressure Hydrothermal Method,* Mat. Res. Bull. **24** (1989), 967.

13. J. E. Geusic, M. Peter, and E. O. Schultz-Dubois, *Paramagnetic Resonance Spectrum of Cr^{+++} in Emerald,* Bell Syst. Tech. J. **38** (1959), 291.

14. Z. Hasan, S. T. Keany, and N. B. Manson, *Spectral Energy Transfer and Fluorescence Line Narrowing in Emerald,* J. Phys. **C19** (1986), 6381.

15. P. Kisliuk and C. A. Moore, *Radiation from the $^4T_2$ State of Cr^{3+} in Ruby and Emerald,* Phys. Rev. **160** (1967), 307.

16. S. T. Lai and B. H. T. Chai, *Spectroscopic and Laser Characterization of Emerald,* Army Research Office Report, ARO-20240 (August 1986), NTIS AD-A172 871.

17. A. D. Liehr, *The Three Electron (or Hole) Cubic Ligand Field Spectrum,* J. Phys. Chem. **67** (1963), 1314.

18. R. M. Macfarlane, *Perturbation Methods in the Calculation of Zeeman Interactions and Magnetic Dipole Line Strengths for d^3 Trigonal-Crystal Spectra,* Phys. Rev. **B1** (1970), 989.

19. R. M. Martirosyan, M. O. Manvelyan, and V. G. Badalyan, *Spin-Lattice Coupling Constants and Spin-Lattice Relaxation for Cr^{3+} Ion in Beryl,* Magn. Reson. Relat. Phenom., Proc. 20th Congr. AMPERE, E. Kundla, E. Lippman, and T. Saluveri, eds., p 318.

20. R. M. Martirosyan, M. O. Manvelyan, G. A. Mnatsakanyan, and V. S. Sevastyanov, *Spin-Lattice Relaxation of Cr^{3+} Ions in Emerald,* Sov. Phys. Solid State **22** (1980), 563.

21. Z. G. Mazurak, M. B. Czaja, J. Hanuza, and B. Jezowska-Trzebiatowska, *The Spectroscopy of Cr^{3+} Doped Natural Garnets and Emerald as well as Synthetic Alexandrite and Corundum,* in Proc. First Int. School on Excited States of Transition Elements, B. Jezowska-Trzebiatowska, J. Legendzie-Wicz, and W. Strek, eds., World Scientific, New Jersey (1989), p 331.

22. T. Miyata, Y. Kawabata, I. Takeda, and H. Kojima, *Habit Modifications of Beryl Grown by the Flux Method,* J. Cryst. Growth **99** (1990), 869.

23. K. Nassau and D. L. Wood, *An Examination of Red Beryl from Utah,* Am. Mineral. **53** (1968), 801.

24. L. V. Nikol'skaya, and M. I. Samoilovich, *Optical Absorption Spectra of Beryls in the Near Infrared (900–2500 nm),* Sov. Phys. Crystallogr. **24** (1979), 604.

25. G. J. Quarles, A. Suchocki, R. C. Powell, and S. Lai, *Optical Spectroscopy and Four-Wave Mixing in Emerald,* Phys. Rev. **B38** (1988), 9996.

26. K. Schmetzer and H. H. Eysel, *Absorption and Emission Spectra of V^{2+}/V^{3+} Doped Beryls,* Z. Naturforsch. **29a** (1974), 1458.

27. P. C. Schmidt, A. Weiss, and T. P. Das, *Effect of Crystal Fields and Self-Consistency on Dipole and Quadrupole Polarizabilities of Closed-Shell Ions*, Phys. Rev. **B19** (1979), 5525.

28. B. V. Shul'gin, M. V. Vasilenko, V. P. Palvanov, and A. V. Kruzhalov, *Electronic Spectra and Structure of Beryl and Chrysoberyl*, Zh. Prik. Spektrosk. **34** (1981), 116.

29. V. P. Solntsev, A. S. Lebedev, V. S. Pavlyuchenko, and V. A. Klyakhin, *Copper Centers in Synthetic Beryl*, Sov. Phys. Solid State **18** (1976), 805.

30. M. V. Vasilenko, A. V. Kruzhalov, and G. V. Bukin, *Excitation Spectra of Synthetic Emerald in the Vacuum Ultraviolet Region*, in Proceedings of the All-Union Conference on Physics of Dielectrics and New Areas for Their Use [in Russian], Karaganda (1978), p 97.

31. D. L. Wood, *Absorption, Fluorescence, and Zeeman Effect in Emerald*, J. Chem. Phys. **42** (1965), 3404.

32. D. L. Wood and K. Nassau, *The Characterization of Beryl and Emerald by Visible and Infrared Absorption Spectroscopy*, Amer. Mineral. **53** (1968), 777.

33. D. L. Wood and K. Nassau, *Infrared Spectra of Foreign Molecules in Beryl*, J. Chem. Phys. **47** (1967), 2220.

34. R.W.G. Wyckoff, *Crystal Structures*, vol 4, Interscience, New York (1968), p 277.

35. W.-C. Zheng, *Spin-Lattice Coupling Coefficients F_{ij}, for d^8 Ions in Cubic Symmetry*, Phys. Status Solidi **(b)159** (1990), K67.

12. $Na_3M_2Li_3F_{12}$ (Fluoride Garnets)

12.1 Crystallographic Data on $Na_3M_2Li_3F_{12}$

Cubic O_h^{10} ($Ia3d$), 230, Z = 8

Ion	Site	Symmetry	x	y	z	q	α (Å^3)
M	16(a)	C_{3i}	0	0	0	3	α_m
Na	24(e)	D_2	0	1/4	1/8	1	0.147^a
Li	24(d)	S_4	0	1/4	3/8	1	0.0321^a
F	96(f)	C_1	x	y	z	−1	0.731^a

^{a}Schmidt et al (1979).

12.2 X-Ray Data on $Na_3M_2Li_3F_{12}$

M	α (Å)	x	y	z	α_m (Å^3)	Ref
Al	12.122	−0.02888	0.04268	0.13989	0.0530	22
Sc	12.607	−0.0343	0.0499	0.1407	0.0540	13
In	12.693	−0.0349	0.0507	0.1422	0.574	13
Ti	12.498	−0.035	0.050	0.140	0.33^a	14
V	12.409	−0.035	0.050	0.140	0.31^a	14
Cr	12.328	−0.035	0.050	0.140	0.29^a	14
Mn	12.141	—	—	—	0.27^a	—
Fe	12.393	−0.035	0.050	0.140	0.24^a	14
Fe	12.387	−0.02954	0.04737	0.14538	0.24^a	10
Co	12.326	−0.035	0.050	0.140	0.23^a	14
Ni	12.446	—	—	—	0.22^a	—
Rh	12.415	−0.035	0.050	0.140	0.71^a	14

^{a}Fraga et al (1976).

12.3 Crystal-Field Data

12.3.1 Crystal-field components, A_{nm} (cm^{-1}/Å^n), for M (C_{3i}) site in $Na_3M_2Li_3F_{12}$ (rotated so that z-axis is along (111) crystallographic axis)

M	Monopole			Total		
	A_{20}	A_{40}	A_{43}	A_{20}	A_{40}	A_{43}
Al	−2051	−14,470	16,491	732.8	−15,454	18,471
Sc	107.7	−10,058	11,840	−307.3	−11,064	13,385
In	−123.3	−9,150	10,803	−482.5	−10,065	12,202
Ti	−466.9	−10,590	12,596	−841.2	−11,667	14,241
V	−477.0	−10,975	13,054	−881.1	−12,090	14,770
Cr	−486.5	−11,340	13,489	−919.6	−12,492	15,273
Fe	−478.8	−11,046	13,139	−888.6	−12,168	14,868
Fe	−3026	−10,404	11,650	1966.2	−11,351	13,247
Co	−486.7	−11,349	13,500	−920.6	−12,502	15,285
Rh	−476.3	−10,948	13,023	−878.4	−12,062	14,734

12.3.2 Theoretical crystal-field parameters, B_{nm} (cm^{-1}), for Al (C_{3i}) site of $Na_3Al_2Li_3F_{12}$ for transition-metal ions with electronic configuration $3d^N$

X^{3+}	N	(a) For monopole A_{nm}			(b) For total A_{nm}		
		B_{20}	B_{40}	B_{43}	B_{20}	B_{40}	B_{43}
Ti	1	−1632	−18,536	21,125	583	−19,797	23,662
V	2	−1372	−13,233	15,081	490	−14,133	16,892
Cr	3	−1185	−10,000	11,397	423	−10,680	12,766
Mn	4	−1037	−7,760	8,844	370	−8,288	9,906
Fe	5	−910	−6,044	6,888	325	−6,455	7,716
Co	6	−824	−5,073	5,782	295	−5,418	6,476
Ni	7	−744	−4,201	4,787	266	−4,486	5,362
Cu	8	−673	−3,492	3,979	240	−3,729	4,457
Zn	9	−611	−2,917	3,325	218	−3,116	3,724

12.3.3 Theoretical crystal-field parameters, B_{nm} (cm^{-1}), for Sc (C_{3i}) site of $Na_3Sc_2Li_3F_{12}$ for triply ionized transition-metal ions with electronic configuration $3d^N$

		(a) For monopole A_{nm}			(b) For total A_{nm}		
X^{3+}	N	B_{20}	B_{40}	B_{43}	B_{20}	B_{40}	B_{43}
Ti	1	−86	13,035	−15,177	−245	−14,173	17,146
V	2	−72	9,306	−10,835	−206	−10,118	12,241
Cr	3	−62	7,033	−8,188	−177	−7,646	9,250
Mn	4	−54	5,457	−6,354	−155	−5,934	7,178
Fe	5	−48	4,251	−4,949	−136	−4,621	5,591
Co	6	−43	3,568	−4,154	−124	−3,879	4,693
Ni	7	−39	2,954	−3,440	−111	−3,212	3,886
Cu	8	−35	2,456	−2,859	−101	−2,670	3,230
Zn	9	−32	2,052	−2,389	−91	−2,231	2,698

12.3.4 Theoretical crystal-field parameters, B_{nm} (cm^{-1}), for In (C_{3i}) site of $Na_3In_2Li_3F_{12}$ for transition-metal ions with electronic configuration $3d^N$

		(a) For monopole A_{nm}			(b) For total A_{nm}		
X^{3+}	N	B_{20}	B_{40}	B_{43}	B_{20}	B_{40}	B_{43}
Ti	1	−98	−11,722	13,839	−384	−12,893	15,631
V	2	−82	−8,368	9,879	−323	−9,204	11,159
Cr	3	−71	−6,324	7,466	−279	−6,956	8,433
Mn	4	−62	−4,907	5,794	−244	−5,398	6,544
Fe	5	−55	−3,822	4,512	−214	−4,204	5,097
Co	6	−50	−3,208	3,788	−194	−3,529	4,278
Ni	7	−45	−2,656	3,136	−175	−2,922	3,542
Cu	8	−40	−2,208	2,607	−158	−2,429	2,944
Zn	9	−37	−1,845	2,178	−144	−2,029	2,460

12.4 Experimental Parameters (cm^{-1}) for Cr^{3+} in $Na_3M_2Li_3F_{12}$

M	$F^{(2)}$	$F^{(4)}$	B_{20}	B_{40}	B_{43}	Ref
In	59,696	42,714	0	−22,218^a	—	5
Ga	59,973	42,242	0	−22,988^a	—	1

^{a}Cubic approximation $B_{43} = \sqrt{10/7}\, B_{40}$

12.5   Bibliography and References

1. J. A. Caird, S. A. Payne, P. R. Staver, A. J. Ramponi, L. L. Chase, and W. F. Krupke, *Quantum Electronic Properties of the $Na_3Ga_2Li_3F_{12}$:Cr^{3+} Laser,* IEEE J. Quantum Electron. **QE-24** (1988), 1077.

2. J. A. Caird, P. R. Staver, M. D. Shinn, H. J. Guggenheim, and D. Bahnck, *Single Crystal $Na_3Ga_2Li_3F_{12}$:Cr^{3+} Laser Pumped Laser Experiments,* in *Advances in Laser Science—II,* M. Lapp, W. C. Stwalley, and G. A. Kenny-Wallace, eds., New York: AIP (1987), pp 120–123.

3. J. Chenavas, J. C. Joubert, M. Marezio, and B. Ferrand, *On the Crystal Symmetry of the Garnet Structure,* J. Less-Common Metals **62** (1978), 373.

4. J. –C. Cretenet, *Composés définis et solutions solides dans des systèmes $M_3^I VF_6$-$M_3^I VF_6$,* C. R. Acad. Sci. **C268** (1969), 2092.

5. D. De Viry, J. P. Denis, B. Blanzat, and J. Grannec, *Spectroscopic Properties of Trivalent Chromium in the Fluoride Garnet $Na_3In_2Li_3F_{12}$,* J. Solid State Chem. **71** (1987), 109.

6. S. Fraga, K.M.S. Saxena, and J. Karwowski, *Handbook of Atomic Data,* Elsevier, New York (1976).

7. N. M. Javoronkov, J. A. Buslaev, and V. P. Tarasov, *Une Etude par résonance magnetique nucléaire du grenat fluoré d'aluminium,* Bull. Soc. Chim. Fr. **7** (1969), 2333.

8. R. H. Langley and G. D. Sturgeon, *Infrared Spectra of Fluoride Garnets,* Spectrochim. Acta **35A** (1979), 209.

9. R. H. Langley and G. D. Sturgeon, *Synthesis of Transition-Metal Fluoride Garnets,* J. Fluorine Chem. **13** (1979), 1.

10. W. Massa, B. Post, and D. Babel, *Verfeinrung der Granatstruktur des Natrium-Lithium-Eisen (III) Fluorids $Na_3Li_3Fe_2F_{12}$,* Z. Kristallogr. **158** (1982), 299.

11. S. Naka, Y. Takeda, K. Kawada, and M. Inagaki, *Growth of Single Crystals of $Na_3Al_2Li_3F_{12}$ Garnet Under Hydrothermal Conditions,* J. Crystal Growth **46** (1979), 461.

12. S. Naka, Y. Takeda, M. Sone, and Y. Suwa, *Synthesis of the Fluoride Garnet {Na_3}(Li_3)F_{12} (M = Al, Cr, Fe),* private communication.

13. R. de Pape, J. Portier, G. Gauthier, and P. Hagenmuller, *Les grenats et des éléments de transition $Na_3Li_3M_2F_{12}$ (M = Ti, V, Cr, Fe ou Co),* C. R. Acad. Sci. **C265** (1967), 1244.

14. R. de Pape, J. Portier, J. Grannec, G. Gauthier, and P. Hagenmuller, *Sur quelques nouveaux grenats fluorés,* C. R. Acad. Sci. **C269** (1969), 1120.

15. H. Pauly, *Gladstone-Dale Calculations Applied to Fluorides,* Can. Mineral. **20** (1982), 593.

16. H. Pauly, *Cryolite, Chiolite, and Cryolithionite: Optical Data Redetermined,* Bull. Geol. Soc. Den. **26** (1977), 95.

17. J. Sawicki and S. S. Hafner, *Sign of Electric Field Gradient at ^{57}Fe Nuclei in Garnets,* Phys. Lett. **68A** (1978), 80.

18. P. C. Schmidt, A. Weiss, and T. P. Das, *Effect of Crystal Fields and Self-Consistency on Dipole and Quadrupole Polarizabilities of Closed-Shell Ions,* Phys. Rev. **B19** (1979), 5525.

19. Y. Takeda, M. Inagaki, and S. Naka, *High Pressure Form of Fluoride Garnets $Na_3M_2Li_3F_{12}$ (M = Al & Fe),* Mater. Res. Bull. **12** (1977), 689.

20. Y. Takeda, M. Sone, Y. Suwa, M. Inagaki, and S. Naka, *Synthesis of Fluoride Garnets $\{Na_3\}[M_2^{3+}Li_3F_{12}]M = Al, Cr, and Fe) from Aqueous Solution and Their Properties,* J. Solid State Chem. **20** (1977), 261.

21. D. de Viry, J. P. Denis, N. Tercier, and B. Blanzat, *Effect of Pressure on Trivalent Chromium Photoluminescence in Fluoride Garnet $Na_3In_2Li_3F_{12}$,* Solid State Commun. **63** (1987), 1183.

22. R.W.G. Wyckoff, *Crystal Structures,* vol 3, Interscience, New York (1968), p 222.

13. Cs₂SnBr₆

13.1 Crystallographic Data on Cs_2SnBr_6

Cubic $O_h^5 (Fm2m)$, 225, Z = 4

Ion	Site	Symmetry	x^a	y	z	q	α (Å³)
Sn	4(a)	O_h	0	0	0	4	0.37^b
Cs	8(c)	T_d	1/4	1/4	1/4	1	2.492^c
Br	24(e)	C_{4v}	x	0	0	−1	3.263^c

^{a}X-ray data: a = 10.81 Å, x = 0.245 (Wyckoff, 1968).
^{b}Fraga et al (1976).
^{c}Schmidt et al (1979).

13.2 Crystal-Field Components, A_{nm} (cm⁻¹/Åⁿ), for Sn (O_h) Site

$A_{nm}{}^a$	Monopole	Self-induced	Dipole	Total
A_{40}	3325	−2256	5088	6157

$^a A_{44} = \sqrt{5/14}\ A_{40}$

13.3 Experimental Parameters (cm⁻¹)

Ion	nd^N	$F^{(2)}$	$F^{(4)}$	ζ	$B_{40}{}^a$	Ref
Os⁴⁺	$5d^4$	88,308	34,676	3212	10,096	5

$^a B_{44} = \sqrt{5/14}\ B_{40}$

13.4 Bibliography and References

1. J. C. Collingwood, P. N. Schatz, and P. J. McCarthy, *Absorption and Magnetic Circular Dichroism Spectra of Ru⁴⁺ in Cs₂ZrCl₆ and Cs₂SnBr₆*, Mol. Phys. **30** (1975), 269.

2. P. B. Dorain, *Magnetic and Optical Properties of Transition Metal Ions in Single Crystals,* Aerospace Laboratories Report, ARL-73-0139 (October 1973), NTIS AD 769870.

3. C. D. Flint, *Luminescence Spectra and Relaxation Process of ReCl₆²⁻ in Cubic Crystals,* J. Chem. Soc. Faraday Trans. 2 **74** (1978), 767.

4. C. D. Flint and A. G. Paulusz, *High Resolution Infrared and Visible Luminescence Spectra of ReCl₆²⁻ and ReBr₆²⁻ in Cubic Crystals,* Mol. Phys. **43** (1981), 321.

5. C. D. Flint and A. G. Paulusz, *High Resolution Infrared and Visible Luminescence Spectra of OsCl₆²⁻ and OsBr₆²⁻ in Cubic Crystals,* Mol. Phys. **41** (1980), 907.

6. S. Fraga, K.M.S. Saxena, and J. Karwowski, *Handbook of Atomic Data,* Elsevier, New York (1976).

7. P. C. Schmidt, A. Weiss, and T. P. Das, *Effect of Crystal Fields and Self-Consistency on Dipole and Quadrupole Polarizabilities of Closed-Shell Ions,* Phys. Rev. **B19** (1979), 5525.

8. R.W.G. Wyckoff, *Crystal Structures,* vol 3, Interscience, New York (1968), p 339.

14. $KMgF_3$

14.1 Crystallographic Data on $KMgF_3$

Cubic O_h^1 $(Pm3m)$, 221, $Z = 1$

Ion	Site	Symmetry	x^a	y	z	q	$\alpha\,(\text{Å}^3)^b$
K	1(a)	O_h	0	0	0	1	0.827
Mg	1(b)	O_h	1/2	1/2	1/2	2	0.0809
F	3(d)	D_{4h}	1/2	1/2	0	−1	0.731

[a]X-ray data: $a = 3.973$ Å (Wyckoff, 1964).
[b]Schmidt et al (1979).

14.2 Crystal-Field Components, A_{nm} (cm^{-1}/Å^n), for Mg (O_h) (1b) Site

$A_{nm}{}^a$	Point charge	Self-induced	Dipole	Total
A_{40}	13,281	−5022	0	8260

[a]$A_{44} = \sqrt{5/14}\ A_{40}$

14.3 Experimental Parameters (cm^{-1})

Ion	nd^N	$F^{(2)}$	$F^{(4)}$	ζ	α	$B_{40}{}^a$	Ref
V^{2+}	$3d^3$	53,362	28,065	—	79	25,515	37
V^{2+}	$3d^3$	49,508	34,871	—	—	22,911	21
Mn^{2+}	$3d^5$	64,099	38,983	—	—	18,659	37
Mn^{2+}	$3d^5$	66,106	39,307	—	—	18,466	17
Mn^{2+}	$3d^5$	59,465	40,446	—	—	18,417	25
Co^{2+}	$3d^7$	70,224	48,787	500	—	16,800	18
Co^{2+}	$3d^7$	71,972	50,001	—	—	14,007	17
Ni^{2+}	$3d^8$	74,480	50,274	620	—	14,658	4
Ni^{2+}	$3d^8$	76,405	53,298	—	—	15,225	14
Ni^{2+}	$3d^8$	74,270	49,322	—	—	15,909	(b)

[a]$B_{44} = \sqrt{5/14}\ B_{40}$
[b]Best fit to the data (Walker and Tang, 1981) with $F^{(2)}$, $F^{(4)}$, and B_{40} varied.

14.4 Bibliography and References

1. M. T. Barriuso and M. Moreno, *Determination of the Mn^{2+}–F^- Distance from the Isotropic Superhyperfine Constant for $[MnF_6]^{4-}$ in Ionic Lattices*, Phys. Rev. **B29** (1984), 3623.

2. J. Brynestad, H. L. Yakel, and G. P. Smith, *Temperature Dependence of the Absorption Spectrum of Nickel (II)-Doped $KMgCl_3$ and the Crystal Structure of $KMgCl_3$*, J. Chem. Phys. **45** (1966), 4652.

3. J. Ferguson, H. J. Guggenheim, L. F. Johnson, and H. Kamimura, *Magnetic Dipole Character of $^3A_{2g} \rightarrow ^3T_{2g}$ Transition in Octahedral Nickel (II) Compounds*, J. Chem. Phys. **38** (1963), 2579.

4. J. Ferguson, H. J. Guggenheim, and D. L. Wood, *Electronic Absorption Spectrum of Ni II in Cubic Perovskite Fluorides: I*, J. Chem. Phys. **40** (1964), 822.

5. J. Ferguson, H. Masui, and Y. Tanabe, *Optical Spectra of Nearest Neighbour Nickel Pairs in KMgF$_3$ and KZnF$_3$ Crystals*, Mol. Phys. **37** (1979), 737.

6. J. Ferguson, D. L. Wood, and K. Knox, *Crystal-Field Spectra of d^3, d^7 Ions: II.—KCoF$_3$, CoCl$_2$, CoBr$_2$ and CoWO$_4$*, J. Chem. Phys. **39** (1963), 881.

7. W. Flassak and H. J. Paus, *Metal-Color Center Complexes in KMgF$_3$:Cu. A New Laser Center*, Cryst. Latt. Def. Amorph. Mater. **15** (1987), 151.

8. J. C. Gacon, A. Gros, H. Bill, and J. P. Wicky, *New Measurements of the Emission Spectra of Sm^{2+} in KMgF$_3$ and NaMgF$_3$ Crystals*, J. Phys. Chem. Solids **42** (1981), 587.

9. T.P.P. Hall, W. Hayes, R.W.H. Stevenson, and J. Wilkens, *Investigation of the Bonding of Iron-Group Ions in Fluoride Crystals: I*, J. Chem. Phys. **38** (1963), 1977.

10. T.P.P. Hall, W. Hayes, R.W.H. Stevenson, and J. Wilkens, *Investigation of the Bonding of Iron-Group Ions in Fluoride Crystals: II*, J. Chem. Phys. **39** (1963), 35.

11. L. F. Johnson, H. J. Guggenheim, and D. Bahnck, *Phonon-Terminated Laser Emission from Ni^{2+} Ions in KMgF$_3$*, Opt. Lett. **8** (1983), 371.

12. L. F. Johnson, H. J. Guggenheim, and R. A. Thomas, *Phonon-Terminated Optical Masers*, Phys. Rev. **149** (1966).

13. P. J. King and J. Monk, *Anomalous Acoustic Relaxation Absorption and Acoustic Paramagnetic Resonance in KMgF$_3$ Containing Fe^{2+}*, Acta Phys. Slovaca **30** (1980), 11.

14. K. Knox, R. G. Shulman, and S. Sugano, *Covalency Effects in KNiF$_3$: II.— Optical Studies*, Phys. Rev. **130** (1963), 512.

15. B. B. Krichevtsov, P. A. Markovin, S. V. Petrov, and R. V. Pisarev, *Isotropic and Anisotropic Magnetic Refraction of Light in the Antiferromagnets KNiF$_3$ and RbMnF$_3$*, Sov. Phys. JETP **59** (1984), 1316.

16. A. L. Larionov, *Calculation of Electric Field Constants for Tetragonal Cr^{3+} Centres in KMeF$_3$ (Me = Mg, Zn)*, Magn. Reson. Relat. Phenom., Proc. Congr., E. Kundla, E. Lippmaa, and T. Saluvere, eds. (1979), p 204.

17. K. H. Lee and W. A. Sibley, *Exchange Enhancement of Co^{2+} and Mn^{2+} Transitions Due to Radiation Defects*, Phys. Rev. **B12** (1975), 3392.

18. A. D. Liehr, *The Three Electron (or Hole) Cubic Ligand Field Spectrum*, J. Phys. Chem. **67** (1963), 1314.

19.   W.L.W. Ludekens and A.J.E. Welch, *Reaction Between Metal Oxides and Fluorides: Some New Double-Fluoride Structures of Type ABF$_3$*, Acta Crystallogr. **5** (1952), 841.

20.   V. Ya. Mitrofanov and A. E. Nikiforov, *Exchange Interactions in the Excited $^3A_2{}^3T_2$ State of Nickel-Ion Pairs in KMgF$_3$*, Opt. Spectrosc. **68** (1990), 508.

21.   R. Moncorge and T. Benyattou, *Excited-State-Absorption and Laser Parameters of V^{2+} in MgF$_2$ and KMgF$_3$*, Phys. Rev. **B37** (1988a), 9177.

22.   R. Moncorge and T. Benyattou, *Excited-State Absorption of Ni^{2+} in MgF$_2$ and MgO*, Phys. Rev. **B37** (1988b), 9186.

23.   S. Muramatsu, *Intensity of the Zero-Phonon Lines for the $^4T_1 \rightarrow {}^4T_2$ Transition of Co^{2+} in KMgF$_3$*, Phys. Status Solidi **(b)98** (1980), K167.

24.   D. E. Onopko, N. V. Starostin, and S. A. Titov, *Interpretation of High-Energy Absorption Spectra of 3d Transition Elements in a KMgF$_3$ Crystal*, Sov. Phys. Solid State **19** (1977), 340.

25.   B. Ng and D. J. Newman, *A Linear Model of Crystal-Field Correlation Effects in Mn^{2+}*, J. Chem. Phys. **84** (1986), 3291.

26.   H. Onuki, F. Sugawara, M. Hirano, and Y. Yamaguchi, *Ultraviolet Photoemission Study of Perovskite Fluorides KMF$_3$ (M:Mn, Fe, Co, Ni, Cu, Zn) in the Valence Band Region*, J. Phys. Soc. Jpn. **49** (1980), 2314.

27.   S. A. Payne and L. L. Chase, *Excited State Absorption of V^{2+} and Cr^{3+} Ions in Crystal Hosts*, J. Lumin. **38** (1987), 187.

28.   S. A. Payne, L. L. Chase, and G. D. Wilke, *Excited-State Absorption Spectra of V^{2+} in KMgF$_3$ and MgF$_2$*, Phys. Rev. **B37** (1988), 998.

29.   A. Poirer and D. Walsh, *Photoluminescence of Iron-Doped KMgF$_3$*, J. Phys. **C16** (1983), 2619.

30.   M. L. Reynolds and G.F.J. Garlick, *The Infrared Emission of Nickel Ion Impurity Centres in Various Solids*, Infrared Phys. **7** (1967), 151.

31.   F. Rodrequez, M. Moreno, A. Tressaud, and J. P. Chaminade, *Mn^{2+} in Cubic Perovskites: Determination of the Mn^{2+}–F$^-$ Distance from the Optical Spectrum*, Cryst. Latt. Def. Amorph. Mater. **16** (1987), 221.

32.   D. K. Sardar, W. A. Sibley, and R. Alcala, *Optical Absorption and Emission from Irradiated RbMgF$_3$:Eu^{2+} and KMgF$_3$:Eu^{2+}*, J. Lumin. **27** (1982), 401.

33.   P. C. Schmidt, A. Weiss, and T. P. Das, *Effect of Crystal Fields and Self-Consistency on Dipole and Quadrupole Polarizabilities of Closed-Shell Ions*, Phys. Rev. **B19** (1979), 5525.

34.   W. A. Sibley, S. I. Yun, and L. N. Feuerhelm, *Radiation Defect and 3d Impurity Absorption in MgF$_2$ and KMgF$_3$ Crystals*, J. Phys. Paris **34** (1973), C9-503.

35.   W. A. Sibley, S. I. Yun, and W. E. Vehse, *Colour Centre Luminescence in KMgF$_3$:Mn Crystals*, J. Phys. **C6** (1973), 1105.

36.   M. D. Sturge, *Dynamic Jahn-Teller Effect in the $^4T_2$ Excited States of $d^{3,7}$ Ions in Cubic Crystals: I.—V^{2+} in KMgF$_3$*, Phys. Rev. **B1** (1970), 1005.

37.   M. D. Sturge, F. R. Merritt, L. F. Johnson, H. J. Guggenheim, and J. P. van der Ziel, *Optical and Microwave Studies of Divalent Vanadium in Octahedral Fluoride Coordination*, J. Chem. Phys. **54** (1971), 405.

38.   E. G. Valyashko, S. N. Bodrug, A. V. Krutikov, V. N. Mednikova, and V. A. Smirnov, *Absorption Spectra of Sm^{2+} in the Double Fluoroperovskites KMgF$_3$ and NaMgF$_3$*, Opt. Spectrosc. (USSR) **44** (1978), 425.

39.   P. J. Walker and H. G. Tang, *Growth of $KMg_{1-x}M_xF_3$ ($M = V^{2+}, Co^{2+}, Ni^{2+}$) Mixed Crystals by the Czochralski Method*, J. Crystal Growth **55** (1981), 539.

40.   D. Walsh and J. Lange, *Dynamic Jahn-Teller Effects in the Photoluminescence of Fe^{2+} in KMgF$_3$*, Phys. Rev. **B23** (1981), 8.

41.   R.W.G. Wyckoff, *Crystal Structures*, vol 2 (1964), p 392.

42.   S. J. Yun, L. A. Kappers, and W. A. Sibley, *Enhancement of Impurity-Ion Absorption Due to Radiation-Produced Defects*, Phys. Rev. **B8** (1973), 773.

43.   S. J. Yun, K. H. Lee, W. A. Sibley, and W. E. Vehse, *Use of 3d-Impurity-Ion Absorption to Study the Distribution of Radiation Damage in Crystals*, Phys. Rev. **B10** (1974), 1665.

44.   E. Zorita, P. J. Alonso, R. Alcala, J. M. Spaeth, and H. Soethe, *Endor Study of Ni^{2+} in KMgF$_3$*, Solid State Commun. **66** (1988), 773.

15. BeAl$_2$O$_4$ (Chrysoberyl, Cr:BeAl$_2$O$_4$ = Alexandrite)

15.1 Crystallographic Data on BeAl$_2$O$_4$

Orthorhombic D_{2h}^{16} (*Pnma*), 62, $Z = 4$

Ion	Site	Symmetry	x	y	z	q	α (Å^3)[a]
Al$_1$	4(a)	C_i	0	0	0	3	0.0530
Al$_2$	4(c)	C_s	x	1/4	z	3	0.0530
Be	4(c)	C_s	x	1/4	z	2	0.0125
O$_1$	4(c)	C_s	x	1/4	z	−2	1.349
O$_2$	4(c)	C_s	x	1/4	z	−2	1.349
O$_3$	8(d)	C_1	x	y	z	−2	1.349

[a]*Schmidt et al (1979)*.

15.2 X-Ray Data

Cell size			Al$_2$		Be	
a	b	c	x	z	x	z
9.4041	5.4756	4.4267	0.27319	−0.00595	0.09294	0.43347
9.407	5.4781	4.4285	0.27283	−0.00506	0.09276	0.43402

O$_1$		O$_2$		O$_3$			
x	z	x	z	x	y	z	Ref
0.09051	0.79016	0.43343	0.24097	0.16318	0.01718	0.25850	42
0.09031	0.78779	0.43302	0.24137	0.16330	0.01529	0.25687	4

15.3 Crystal-Field Components

A_{nm} (cm^{-1}/Å^n) for Al_2 (C_s) site (rotated so that z-axis is perpendicular to mirror plane)

A_{nm}	Calculated using data from Wyckoff (1968)		A_{nm}	Calculated using data from Dudka et al (1985)	
	Monopole	Total		Monopole	Total
ReA_{11}	−1,474	−2533	ReA_{11}	−1,247	−2303
ImA_{11}	−5,753	−1586	ImA_{11}	−6,656	−1410
A_{20}	4,776	−658.4	A_{20}	4,828	−471.4
A_{22}	5,573	6129	A_{22}	5,584	5661
ReA_{31}	6,459	3852	ReA_{31}	6,754	3340
ImA_{31}	−7,028	3936	ImA_{31}	−6,192	3415
ReA_{33}	−4,503	−346.7	ReA_{33}	−4,408	−79.97
ImA_{33}	−1,662	46.55	ImA_{33}	−2,256	−50.58
A_{40}	−4,588	−1744	A_{40}	−4,872	−1994
ReA_{42}	−18,247	−2571	ReA_{42}	−17,366	−2231
ImA_{42}	−12,176	8657	ImA_{42}	−13,550	8786
ReA_{44}	4,922	−3188	ReA_{44}	3,246	−3544
ImA_{44}	11,144	−4138	ImA_{44}	11,736	−3798
ReA_{51}	180.6	322.5	ReA_{51}	153.9	378.6
ImA_{51}	−2,189	1879	ImA_{51}	−1,991	1914
ReA_{53}	−1,037	1287	ReA_{53}	−969.5	1295
ImA_{53}	−1,271	−687.8	ImA_{53}	−1,395	−849.1
ReA_{55}	−1,674	812.1	ReA_{55}	−1,566	984.6
ImA_{55}	360.7	1159	ImA_{55}	4,210	994.5

15.4 Experimental Values (cm^{-1}) of $F^{(2)}$, $F^{(4)}$, and B_{40} for nd^N Ions

Ion	d^N	$F^{(2)}$	$F^{(4)}$	B_{40}[a]	Ref
Cr^{3+}	$3d^3$	58,765[b]	41,391	35,280	17
Ti^{3+}	$3d^1$	−	−	40,320	20
Ni^{2+}	$3d^8$	71,862	50,501	19,761	20

[a]Cubic approximation $B_{44} = \sqrt{5/14} \ B_{40}$, C_2 site.
[b]Only B given $C/B = 4.5$ assumed.

15.5 Bibliography and References

1. G. V. Bukin, S. Yu. Volkov, V. N. Matrosov, B. K. Sevast'yanov, and M. I. Timosheshkin, *Stimulated Emission from Alexandrite* ($BeAl_2O_4$:Cr^{3+}) Sov. J. Quantum Electron. **8** (1978), 671.

2. Y. Chiba, K. Yamagishi, and H. Ohkura, *ESR of Ti^{3+} Ions in Laser-Quality Chrysoberyl*, Jpn. J. Appl. Phys. **27** (1988), L1929.

3. C. F. Cline, R. C. Morris, M. Dutoit, and P. J. Harget, *Physical Proper-*
    ties of $BeAl_2O_3$ Single Crystals, J. Mater. Sci. **14** (1979), 941.

4. A. P. Dudka, B. K. Sevast'yanov, and V. I. Simonov, *Refinement of*
    Atomic Structure of Alexandrite, Sov. Phys. Crystallogr. **30** (1985), 277.

5. C. E. Forbes, *Analysis of the Spin-Hamiltonian Parameters for Cr^{3+} in*
    Mirror and Inversion Symmetry Sites of Alexandrite ($Al_{2-x}Cr_xBeO_4$).
    Determination of the Relative Site Occupancy by EPR, J. Chem. Phys. **79**
    (1983), 2590.

6. S. K. Gayen, W. B. Wang, V. Petricevic, and R. R. Alfano, *Picosecond*
    Time Resolved Studies of Nonradiative Relaxation in Ruby and Alexan-
    drite, AIP Conf. Proc. **146** (1986), 206.

7. A. M. Ghazzawi, J. K. Tyminski, R. C. Powell, and J. C. Walling, *Four-*
    Wave Mixing in Alexandrite Crystals, Phys. Rev. **B30** (1984), 7182.

8. Xiang-an Guo, Bang-xing Zhang, Lu-Sheng Wu, and Mei-Ling Chen,
    Czochralski Growth and Laser Performance of Alexandrite Crystals, AIP
    Conf. Proc. **146**, New York (1986), p 249.

9. Guo Xingan, Chen Meiling, Li Nairen, Qin Qinghai, Huang Mingfang,
    Fei Jingwei, Wen Shulin, Li Zongquan, and Qin Yong, *Czochralski*
    Growth of Alexandrite Crystals and Investigation of their Effect, J. Cryst.
    Growth **83** (1987), 311.

10. F. Hasen and A. El-Rakhawy, *Chromium III Centers in Synthetic*
    Alexandrite, Amer. Minerol. **59** (1974), 159.

11. L. A. Harris and H. L. Yakel, *Synthesis and X-Ray Study of Single-*
    Crystal $3Al_2O_3 \rightarrow BeO$, J. Amer. Ceramic Soc. (June 1970), 359.

12. D. F. Heller and J. C. Walling, *High-Power Performance of Alexandrite*
    Lasers, Cleo '84 (20 June 1984), p 101.

13. A. M. Hofmeister, T. C. Hoering, and D. Virgo, *Vibrational Spectros-*
    copy of Beryllium Aluminosilicates: Heat Capacity Calculations from
    Band Assignments, Phys. Chem. Miner. **14** (1987), 205.

14. P. T. Kenyon, L. Andrews, B. McCollum, and A. Lempicki, *Tunable*
    Infrared Solid-State Laser Materials Based on Cr^{3+} in Low Ligand
    Fields, IEEE J. Quantum Electron. **QE-18** (1982), 1189.

15. T. Kottke and F. Williams, *Pressure Dependence of the Alexandrite*
    Emission Spectrum, Phys. Rev. **B28** (1983), 1923.

16. S. Majetich, D. Boye, J. E. Rives, and R. S. Meltzer, *Spectroscopy of the*
    ^2E Excited State of Cr^{3+} in Alexandrite, J. Lumin. **40,41** (1988), 207.

17. Z. G. Mazurak, M. B. Czaja, J. Hanuza, and B. Jezowski-Trzebiotowska,
    The Spectroscopy of Cr^{3+} Doped Natural Garnets and Emerald as well
    as Synthetic Alexandrite and Corundum, Proc. First International School
    on Excited States of Transition Elements, B. Jezowski-Trzebiatowska, J.

Legendziewicz, and W. Strek, eds., World Scientific, New Jersey (1989), p 331.

18. R. E. Newnham, J. J. Kramer, W. A. Schulze, and L. E. Cross, *Magnetoferroelectricity in Cr$_2$BeO$_4$*, J. Appl. Phys. **49** (1978), 6088.

19. S. A. Payne and L. L. Chase, *Excited State Absorption of V^{3+} and Cr^{3+} Ions in Crystal Hosts*, J. Lumin. **38** (1987), 187.

20. E. V. Pestryakov, V. I. Trunov, and A. I. Alimpiev, *Induced Emission from Iron-Group Ions in Chrysoberyl*, Izv. Akad. Nauk SSSR Ser. Fiz. **52** (1988), 1184.

21. E. V. Pestryokov, V. I. Trunov, and A. I. Alimpiev, *Generation of Tunable Radiation in a BeAl$_2$O$_4$:Ti^{3+} Laser Subjected to Pulsed Coherent Pumping at a High Repetition Frequency*, Sov. J. Quantum Electron. **17** (1987), 585.

22. R. C. Powell, Lin Xi, Xu Garg, G. J. Quarles, and J. C. Walling, *Spectroscopic Properties of Alexandrite Crystals*, Phys. Rev. **B32** (1985), 2788.

23. R. C. Sam, W. R. Rapport, and S. Matthews, *New Developments in High-Power High Repetition Rate Injection-Locked Alexandrite Lasers*, Cleo '84 (20 June 1984), p 101.

24. H. Samelson and D. J. Harter, *High-Pressure Arc Lamp Excited cw Alexandrite Lasers*, Cleo '84 (20 June 1984), p 101.

25. K. L. Schepler, *Fluorescence of Inversion Site Cr^{3+} Ions in Alexandrite*, J. Appl. Phys. **56** (1984), 1314.

26. P. Schmidt, A. Weiss, and T. P. Das, *Effect of Crystal Fields and Self-Consistency on Dipole and Quadrupole Polarizabilities of Closed-Shell Ions*, Phys. Rev. **B19** (1979), 5525.

27. Y. Segawa, A. Sugimoto, P. H. Kim, S. Namba, K. Yamagishi, Y. Anzai, and Y. Yamaguchi, *Optical Properties and Lasing of Ti^{3+} Doped BeAl$_2$O$_4$*, Topical Meeting on Tunable Solid State Lasers, vol 26–28 (October 1987), p 154.

28. Y. Segawa, A. Sugimoto, P. H. Kim, S. Namba, K. Yamagishi, Y. Anzai, and Y. Yamaguchi, *Optical Properties and Lasing of Ti^{3+} Doped BeAl$_2$O$_4$*, Jpn. J. Appl. Phys. **26** (1987), L291.

29. B. K. Sevast'yanov, Y. L. Remigailo, V. P. Orekhova, V. P. Matrozov, E. G. Tsvetkov, and G. V. Bukin, *Spectroscopic and Lasing Characteristics of an Alexandrite (Chromium +3) Doped Beryllium Aluminum Oxide Laser*, Dokl. Akad. Nauk SSSR **256** (1981), 373–376; Sov. Phys. Dokl. **26**, No. 1 (January 1981) 62–64.

30. M. L. Shand and H. P. Jenssen, *Energy Kinetics in Alexandrite*, Proc. Int. Conf. Lasers (1983), p 559.

31. M. L. Shand, J. C. Walling, and R. C. Morris, *Excited State Absorption in the Pump Region of Alexandrite,* J. Appl. Phys. **52** (1981), 953.

32. Ma Shiaoshan, Lu Jiajin, Qian Zhenying, Hou Yinchun, Wang Siting, Shen Yajang, and Jing Zonru, *The Growth Habits of Alexandrite Crystal,* Chin. J. Phys. Peking Engl. Transl. **4** (1984), 771.

33. A. Sugimoto, Y. Segawa, P. H. Kim, and S. Namba, *Spectroscopic Properties of Ti^{3+}-Doped $BeAl_2O_4$,* J. Opt. Soc. Am. **B6** (1989), 2334.

34. G. J. Troup, A. Edgar, D. R. Hutton, and P. P. Phakey, *8mm Wavelength EPR Spectrum of Cr^{3+} in Laser-Quality Alexandrite,* Phys. Status Solidi **(a)71** (1982), K29.

35. J. C. Walling, H. P. Jenssen, R. C. Morris, E. W. O'Dell, and O. G. Peterson, *Tunable-Laser Performance in $BeAl_2O_4$:Cr^{3+},* Opt. Lett. **4** (1979), 182.

36. J. C. Walling and O. G. Peterson, *High Gain Laser Performance in Alexandrite,* IEEE J. Quantum Electron. **QE-16** (1980), 119.

37. J. C. Walling and O. G. Peterson, *High Gain Laser Performance in Alexandrite,* IEEE OSA Conf. on Laser Engineering and Applications (1979), p 86D.

38. J. C. Walling, O. G. Peterson, H. P. Jenssen, R. C. Morris, and E. W. O'Dell, *Tunable Alexandrite Lasers,* IEEE J. Quantum Electron. **QE-16** (1980), 1302.

39. J. C. Walling, O. G. Peterson, and R. C. Morris, *Tunable CW Alexandrite Laser,* IEEE J. Quantum Electron. **QE-16** (1980), 120.

40. S. C. Weaver and S. A. Payne, *Determination of Excited-State Polariz-abilities of Cr^{3+}-Doped Materials by Degenerate Four-Wave Mixing,* Phys. Rev. **B40** (1989), 10,727.

41. Wu Guang-Zhao and Zhang Xiu-rong, *Crystal-Field Energy Levels of $BeAl_2O_4$:Cr^{3+},* Acta Phys. Sin. **32** (1983), 64–70 (translation, Chin. J. Phys. Peking Engl. Transl. **3** (1983), 570).

42. R.W.G. Wyckoff, *Crystal Structures,* vol 3 (1965), p 91, and vol 4 (1968), p 159.

43. Xu Jian, Xiong, Guang-nan, and Xu Xu-rong, *Properties of the Excited States of Cr^{3+} in Alexandrite Crystal,* Chin. Sci. Bull. **34** (1989), 199.

44. Zhang Shoudu and Zhang Kemin, *Experiment on Laser Performance of Alexandrite Crystals,* Chin. J. Lasers **11** (1984), 44–46 (translation, Chin. J. Phys. Peking Engl. Transl. **4** (1984), 667).

16. ZnAl₂O₄

16.1 Crystallographic Data on ZnAl₂O₄

Cubic O_h^7 ($Fd3m$), 227, $Z = 8$

Ion	Site	Symmetry	x^a	y	z	q	α (Å³)[b]
Zn	8(a)	T_d	0	0	0	2	0.676
Al	16(d)	D_{3d}	5/8	5/8	5/8	3	0.0530
O	32(e)	C_{3v}	x	x	x	-2	1.349

[a]X-ray data: $a = 8.0883$ Å, $x = 0.390$ (Wyckoff, 1968).
[b]Schmidt et al (1979).

16.2 Crystal Fields for Zn (T_d) Site

16.2.1 Crystal-field components, A_{nm} (cm⁻¹/Åⁿ), for Zn (T_d) site

A_{nm}[a]	Monopole	Dipole	Self-induced	Total
A_{32}	27,467	-7189	-6951	1,327
A_{40}	-12,005	4187	4544	-3,274

[a]$A_{44} = \sqrt{5/14}\ A_{40}$

16.2.2 Crystal-field components, A_{nm} (cm⁻¹/Åⁿ), for Al (D_{3d}) site (rotated so that z-axis is parallel to (111) crystallographic axis)

A_{nm}	Monopole	Dipole	Self-induced	Total
A_{20}	5,004	22,132	-2576	24,559
A_{40}	-21,556	-3,990	9380	16,162
A_{43}	-24,259	-1,899	9322	-16,830

16.3 Experimental Values (cm⁻¹) of $F^{(2)}$, $F^{(4)}$, α, ζ, and B_{nm} for nd^N Ions

Ion	d^N	$F^{(2)}$	$F^{(4)}$	α	ζ	B_{20}	B_{40}	B_{43}	Ref
Co²⁺	3d⁷	59,367	42,210	86	420	—	-8,640[a]	—	9
Cr³⁺	3d³	56,700	40,320	—	250	4608	-30,625	-28,415	3,7,14
Cr³⁺	3d³	56,557	40,370	—	250	5321	-32,312	-28,500	10
Cr³⁺	3d³	56,235	40,144	—	250	3780	-31,661	-28,480	10[b]
Cr³⁺	3d³	56,595	40,396	—	250	3500	-31,290	-28,212	10[b]

[a]$B_{44} = \sqrt{5/14}\ B_{40}$
[b]Nonequivalent centers.

16.4   Bibliography and References

1.   G. Burns, E. A. Geiss, B. A. Jenkins, and M. I. Nathan, *Cr^{3+} Fluorescence in Garnets and Other Crystals,* Phys. Rev. **A139** (1965), 1687.

2.   J. Ferguson and D. L. Wood, *Crystal Field Spectra of d^{3,7} Ions: VI.— The Weak Field Journalism and Covalency,* Aust. J. Chem. **23** (1970), 861.

3.   J. Ferguson, D. L. Wood, and L. G. van Uitert, *Crystal-Field Spectra of d^{3,7} Ions: V.—Tetrahedral Co^{2+} in ZnAl$_2$O$_4$ Spinel,* J. Chem. Phys. **51** (1969), 2904.

4.   R. D. Gillen and R. E. Salomon, *Optical Spectra of Chromium (III), Cobalt (II), and Nickel (II) Ions in Mixed Spinels,* J. Phys. Chem. **74** (1970), 4252.

5.   P. P. Kisluik, D. L. Wood, R. M. Macfarlane, G. F. Imbusch, and D. M. Larkin, *Optical Spectrum of Cr^{3+} Ions in Spinels,* Air Force Report No. TR-68-148 (March 1968), NTIS AD 668446.

6.   R. M. Macfarlane, *Perturbation Methods in the Calculation of Zeeman Interaction and Magnetic Dipole Line Strengths for d^3 Trigonal-Crystal Spectra,* Phys. Rev. **B1** (1970), 989.

7.   W. Mikenda, *N-Lines in the Luminescence Spectra of Cr^{3+}-Doped Spinels (III). Partial Spectra,* J. Lumin. **26** (1981), 85.

8.   W. Mikenda and A. Preisinger, *N-Lines in the Luminescence Spectra of Cr^{3+}-Doped Spinels (I). Identification of N-Lines,* J. Lumin. **26** (1981), 53.

9.   W. Mikenda and A. Preisinger, *N-Lines in the Luminescence Spectra of Cr^{3+}-Doped Spinels (II). Origins of N-Lines,* J. Lumin. **26** (1981), 67.

10. W. Nie, F. M. Michel-Calendini, C. Linares, G. Boulen, and C. Daul, *New Results on Optical Properties and Term-Energy Calculations in Cr^{3+}-Doped ZnAl$_2$O$_4$,* J. Lumin. **46** (1990), 177.

11. P. C. Schmidt, A. Weiss, and T. P. Das, *Effect of Crystal Fields and Self-Consistency on Dipole and Quadrupole Polarizabilities of Closed-Shell Ions,* Phys. Rev. **B19** (1979), 5525.

12. A. van Die, A. Leenaers, W. van der Weg, and G. Blasse, *A Search for Luminescence of the Trivalent Manganese Ion in Solid Aluminates,* Mater. Res. Bull. **22** (1987), 781.

13. D. L. Wood, W. E. Burke, and L. G. van Uittert, *Zeeman Effect of Cr^{3+} in ZnAl$_2$O$_4$,* J. Chem. Phys. **51** (1969), 1966.

14. D. L. Wood, G. F. Imbusch, R. M. Macfarlane, P. Kisliuk, and D. M. Larkin, *Optical Spectrum of Cr^{3+} Ions in Spinels,* J. Chem. Phys. **48** (1968), 5255.

15. R.W.G. Wyckoff, *Crystal Structures,* vol 3, Interscience, New York (1968), p 75.

17. Li$_2$MgZrO$_4$

17.1 Crystallographic Data on Li$_2$MgZrO$_4$

Tetragonal D_{4h}^{19} ($I4_1/amd$), 141 (second setting), $Z = 2$

Ion	Site	Symmetry	x^a	y	z	q	α (Å^3)
O	8(e)	C_{2v}	0	1/4	0.108	−2	1.349
Zr,Mg	4(a)	D_{2d}	0	3/4	1/8	4, 2	0.280
Li	4(b)	D_{2d}	0	1/4	3/8	1	0.0321

^{a}X-ray data: $a = 4.209$ Å, $c = 9.145$ Å (Castellanos et al, 1985).

17.2 Crystal-Field Components, A_{nm} (cm^{-1}/Å^n) for 4(a) (D_{2d}) Site

Assuming that average charge on site 4(a) is +3

A_{nm}	Monopole	Self-induced	Dipole	Total
A_{20}	−4,837	99.2	13,303	8,564
A_{32}	618.0	−854.3	4,448	4,212
A_{40}	19,053	−5338	2,835	16,550
A_{44}	11,142	−3404	−430.3	7,308
A_{52}	−1,763	599.6	−1,229	−2,393

17.3 Bibliography and References

1. M. Castellanos, M. C. Martinez, and A. R. West, *New Family of Phases, Li$_2$MXO$_4$: X = Zr, Hf; M = Mg, Mn, Fe, Co, Ni, Cu, Zn with α-LiFeO$_2$ and Related Structures*, Z. Kristallogr. **190** (1990), 161.

2. M. Castellanos, A. R. West, and W. B. Reid, *Dilithium Magnesium Zirconium Tetraoxide with an α-LiFeO$_2$ Structure*, Acta Crystallogr. **C41** (1985), 1707.

3. V. Danek and K. Matiasovsky, *Preparation of Double Oxides in Ionic Melts*, Z. Anorg. Allg. Chem. **584** (1990), 207.

4. A. Huanosta, M. A. Castellanos, R. Margarita, M. Chavez, and A. R. West, *Dielectric Properties of a New Family of Complex Oxides of the Type Li$_2$ABO$_4$*, Rev. Mex. Fis. **36** (1990), 258 (in Spanish).

18. $La_3Lu_2Ga_3O_{12}$

18.1 Crystallographic Data on $La_3Lu_2Ga_3O_{12}$

Cubic O_h^{10} (Ia3d), 230, Z = 8

Ion	Site	Symmetry	x^a	y	z	q	$\alpha\,(\text{Å}^3)^b$
La	24(c)	D_2	0	1/4	1/8	3	1.41
Lu	16(a)	C_{3i}	0	0	0	3	0.77
Ga	24(d)	S_4	0	1/4	3/8	3	0.458
O	96(h)	C_1	−0.02976	0.05819	0.15699	−2	1.349

^{a}X-ray data: a = 12.93 (Allik et al, 1988).

^{b}Values for α are from Schmidt et al (1979); for values not given there, α values are from Fraga et al (1976).

18.2 Crystal-Field Components, A_{nm} $(\text{cm}^{-1}/\text{Å}^n)$

18.2.1 For Ga ion in 24(d) (S_4) site

A_{nm}	Point charge	Self-induced	Dipole	Total
A_{20}	10,298	−2189	8197	16,306
ReA_{32}	−14,919	4063	−7814	−18,670
ImA_{32}	32,502	−8342	6011	30,171
A_{40}	−18,159	7076	−5524	−16,607
ReA_{44}	−4,696	1758	4006	−2,538
ImA_{44}	−5,156	2214	−2319	−5,260
ReA_{52}	−1,758	1130	−1683	−2,311
ImA_{52}	3,807	−2351	2290	3,475
$|A_{44}|$	6,974	—	—	5,840

18.2.2 For Lu ion in 16(a) (C_{3i}) site (rotated so that z-axis is parallel to (111) crystallographic axis)

A_{nm}	Point charge	Self-induced	Dipole	Total
A_{20}	8,619	−839	−9970	−2190
A_{40}	−10,698	3041	8056	−6852
ReA_{43}	600	−316	2583	2867
ImA_{43}	−11,521	3056	−13	−8478
$|A_{43}|$	11,537	—	—	8950

18.2.3 For La ion in 24(c) (D_2) site[a]

A_{nm}	Point charge	Self-induced	Dipole	Total
A_{20}	−1196	−369	11,910	10,345
A_{22}	1305	−155	−107	43
A_{32}	$-i1029$	$-i34$	$i1,453$	$i390$
A_{40}	-663	−70	−20	753
A_{42}	5073	−990	−323	3,760
A_{44}	−2522	530	971	−1,021
A_{52}	$i1783$	$-i477$	$i106$	$i1,414$
A_{54}	$i967$	$-i253$	$-i87$	$i627$
A_{60}	−1203	336	4	−863
A_{62}	549	−213	200	536
A_{64}	567	−193	−235	139
A_{66}	−474	166	2021	06
A_{72}	$-i73$	$i32$	$i124$	$i83$
A_{74}	$i106$	$-i60$	$i68$	$i114$
A_{76}	$i159$	$-i54$	$-i43$	$i62$

[a] $i = \sqrt{-1}$

18.3 Experimental Parameters (cm^{-1})

Ion	Symmetry	$F^{(2)}$	$F^{(4)}$	B_{40}	T	Ref
Cr^{3+}	C_{3i}	49,830	35,097	−20,720	4 K	10

18.4 Bibliography and References

1. T. H. Allik, S. A. Stewart, D. K. Sardar, G. J. Quarles, R. C. Powell, C. A. Morrison, G. A. Turner, M. R. Kokta, W. W. Hovis, and A. A. Pinto, *Preparation, Structure, and Spectroscopic Properties of Nd^{3+}:{La_{1-x} Lu_x}$_3$[$Lu_{1-y}Ga_y$]$_2Ga_3O_{12}$ Crystals,* Phys. Rev. **B37** (1988), 9129.

2. S. Fraga, J. Karwowski, and K.M.S. Saxena, *Handbook of Atomic Data,* Elsevier, New York (1976), 319.

3. F. M. Hashmi, K. W. VerSteeg, F. Durville, and R. C. Powell, *Four-Wave Mixing Spectroscopy of Cr-Doped Garnet Crystals,* Phys. Rev. **B42** (1990), 3818.

4. V. F. Kitaeva, E. V. Zharikov, and I. L. Christyi, *The Properties of Crystals with Garnet Structure,* Phys. Status Solidi **(a)92** (1985), 475.

5. M. Kokta and M. Grasso, *New Substituted Gallium Garnets Containing Trivalent Lanthanum on Dodecahedral Crystallographic Sites,* J. Solid State Chem. **8** (1973), 357.

6. C. A. Morrison, E. D. Filer, N. P. Barnes, and G. A. Turner, *Theoretical Temperature-Dependent Branching Ratios and Laser Thresholds of the $^5I_7 \rightarrow {}^5I_8$ Levels of Ho^{3+} in Ten Garnets,* Harry Diamond Laboratories, HDL-TR-2185 (September 1990).

7. K. Petermann and G. Huber, *Broad Band Fluorescence of Transition Metal Doped Garnets and Tungstates,* J. Lumin. **31,32** (1984), 71.

8. D. K. Sardar, G. J. Quarles, R. C. Powell, and M. R. Kokta, *Spectroscopic Properties of La$_3$Lu$_2$Ga$_3$O$_{12}$:Nd^{3+} Crystals,* AIP Proc. **146**, New York (1986).

9. P. C. Schmidt, A. Weiss, and T. P. Das, *Effect of Crystal Fields and Self-Consistency on Dipole and Quadrupole Polarizabilities of Closed-Shell Ions,* Phys. Rev. **B19** (1979), 5525.

10. B. Struve and G. Huber, *The Effect of the Crystal Field Strength on the Optical Spectra of Cr^{3+} in Gallium Garnet Laser Crystals,* Appl. Phys. **B36** (1985), 195.

12. E. V. Zharikov, A. S. Zolot'ko, V. F. Kitaeva, V. V. Laptev, V. V. Osiko, N. N. Sobolev, and I. A. Sychev, *Measurement of Elastic and Photoelastic Constants of the Garnet {La$_2$Nd$_{0.3}$Lu$_{0.7}$} Lu$_2$Ga$_3$O$_{12}$,* Sov. Phys. Solid State **25** (1983), 568.

11. M. Yamaga, B. Henderson, K. P. O'Donnell, C. T. Cowan, and A. Marshall, *Temperature Dependence of the Lifetime of Cr^{3+} Luminescence in Garnet Crystals I,* Appl. Phys. **B50** (1990), 425.

19. ZnO

19.1 Crystallographic Data on ZnO

Hexagonal C_{6v}^4 ($P6_3mc$), 186, $Z = 2$

Ion	Site	Symmetry	x	y	z	q	$\alpha\,(\text{Å}^3)^a$
Zn	2(b)	C_{3v}	1/3	2/3	0	2	0.676
O	2(b)	C_{3v}	1/3	2/3	z	-2	1.349

^{a}Schmidt et al (1979).

19.2 X-Ray Data

a	c	z	Ref
3.24950	5.2069	0.345	34
3.24270	5.1948	0.3826	26
3.24986	5.20662	0.3825	1

19.3 Crystal-Field Components, A_{nm} (cm^{-1}/Å^n), for Zn (C_{3v}) Site

Calculated using data from Wyckoff (1968)

A_{nm}	Point charge	Self-induced	Dipole	Total
A_{10}	31,224	0	7,662	38,886
A_{20}	17,809	$-2,962$	22,891	37,738
A_{30}	36,957	$-9,440$	8,892	36,409
A_{33}	15,279	$-3,287$	$-4,742$	7,249
A_{40}	10,043	$-5,353$	4,269	8,959
A_{43}	$-8,681$	2,903	1,141	$-4,637$
A_{50}	4,136	$-3,662$	6,358	6,831
A_{53}	739.7	-578.3	1,254	1,415

Calculated using data from Sabine and Hogg (1969)

A_{nm}	Point charge	Self-induced	Dipole	Total
A_{10}	28,160	0	-311.5	27,849
A_{20}	$-1,685$	254.0	14,628	13,197
A_{30}	29,882	$-6,758$	6,088	29,211
A_{33}	20,084	$-4,709$	$-3,930$	11,446
A_{40}	8,622	$-3,204$	-449.8	4,967
A_{43}	$-9,102$	3,384	-167.6	$-5,885$
A_{50}	-551.3	269.1	2,694	2,411
A_{53}	-886.2	250.6	1,696	1,060

Calculated using data from Abrahams and Bernstein (1969)

A_{nm}	Point charge	Self-induced	Dipole	Total
A_{10}	28,033	0	−293.4	27,740
A_{20}	−1,631	244.7	14,444	13,058
A_{30}	29,635	−6658	5,992	28,968
A_{33}	19,895	−4633	−3,870	11,392
A_{40}	8,522	−3148	−435.8	4,938
A_{43}	−9,003	3325	−161.1	−5,840
A_{50}	−536.4	258.1	2,646	2,368
A_{53}	−868.7	242.5	1,661	1,035

19.4   Experimental Parameters (cm^{-1})

Ion	nd^N	$F^{(2)}$	$F^{(4)}$	ζ	B_{20}	B_{40}	B_{43}	Ref
Co^{2+}	$3d^7$	62,388	43,943	630	—	5460	—	33[a]
Co^{2+}	$3d^7$	61,250	44,100	450	−590	4077	7331	19
Co^{2+}	$3d^7$	56,530	39,690	500	—	5460	—	17[a]
Co^{2+}	$3d^7$	61,740	44,100	430	−785	6665	6248	16
Ni^{2+}	$3d^8$	63,630	46,620	500	96	7072	6529	2
Ni^{2+}	$3d^8$	63,602	46,570	500	—	5880	—	33[a]
Ni^{2+}	$3d^8$	63,218	43,674	630	—	5670	—	23[a]
Cu^{3+}	$3d^9$	—	—	800	—	7000	—	33[a]

[a]Cubic approximation, $B_{20} = 0$, $B_{43} = \sqrt{10/7}\ B_{40}$

19.5   Bibliography and References

1. S. C. Abrahams and J. L. Bernstein, *Remeasurement of the Structure of Hexagonal ZnO,* Acta Crystallogr. **B25** (1969), 1233.

2. R. S. Anderson, *Lattice-Vibration Effects in the Spectra of ZnO:Ni and ZnO:Co,* Phys. Rev. **164** (1967), 398.

3. G. D. Archard, *Anomalous Lattice Constants of Zinc Oxide,* Acta Crystallogr. **6** (1953), 657.

4. W. Bond, *Measurement of Refractive Indices of Several Crystals,* J. Appl. Phys. **36** (1965), 1674.

5. I. J. Broser, R.K.F. Germer, H.-Joachin, E. Shultz, and K. P. Wisznewski, *Fine Structure and Zeeman Effect of the Excited State of the Green Emitting Copper Centre in Zinc Oxide,* Solid-State Electron. **21** (1978), 1597.

6. H. E. Brown, *Zinc Oxide Properties and Applications,* International Lead Zinc Research Organization, New York, NY (1981). (This reference contains a very extensive bibliography.)

7. R. Collins and D. Kleinman, *Infrared Reflectivity of ZnO*, Phys. Chem. Solids **11** (1959), 190.

8. A. Cornet, A. Miralles, O. Ruiz, and J. R. Morante, *Near Infrared Photoluminescence of ZnO:Co Varistors*, Phys. Status Solidi **(a)120** (1990), K105.

9   R. Dingle, *Luminescent Transitions Associated with Divalent Impurities and the Green Emission from Semiconductivity ZnO*, Phys. Rev. Lett. **23** (1969), 579.

10. I. T. Drapak, *Growing Zinc Oxide Single Crystals and Films*, Neorg. Mater. **16** (1980), 362.

11. Du Mao-Lu and Zhao Min-Guang, *Zero-Field Splitting of Tetrahedral Co^{2+} in the Trigonal Crystal Field*, J. Phys. **C21** (1988), 1561.

12. K. Fischer and E. Sinn, *On the Preparation of ZnO Single Crystals*, Cryst. Res. Tech. **16** (1981), 689.

13. G. Heiland, E. Mollwo, and F. Stockmann, *Electronic Processes in Zinc Oxide*, Solid State Phys. **8** (1959), 193.

14. W. Johnston, *Characteristics of Optically Pumped Platelet Lasers of ZnO, CdS, CdSe, and CdS-Se Between 300° and 80°K*, J. Appl. Phys. **42** (1971), 2731.

15. S. A. Kazandzhiev, M. M. Malov, V. D. Chernyi, M. N. Mendakov, A. N. Lobachev, and I. P. Kuzmina, *Effect of Growth Conditions on Optical Properties of Zinc Oxide Single Crystals Grown by Hydrothermal Synthesis*, Opticheshie Issledovanja Polaysrovdnikov (1980), p 99.

16. P. Kiodl, *Optical Absorption of Co^{2+} in ZnO*, Phys. Rev. **B15** (1977), 2493.

17. A. D. Liehr, *The Three Electron (or Hole) Cubic Ligand Field Spectrum*, J. Phys. Chem. **67** (1963), 1314. (This reference contains a multitude of references to earlier work.)

18. D. Louer, J. P. Auffredic, J. I. Langford, D. Ciosmak, and J. C. Niepce, *A Precise Determination of the Shape, Size and Distribution of Size of Crystallites in Zinc Oxide by Oxides by X-Ray Line-Broadening Analysis*, J. Appl. Crystallogr. **16** (1983), 183.

19. R. M. Macfarlane, *Perturbation Methods in the Calculation of Zeeman Interactions and Magnetic Dipole Line Strengths for d^3 Trigonal-Crystal Spectra*, Phys. Rev. **B1** (1970), 989.

20. W. C. Mackrodt, R. F. Stewart, J. C. Campbell, and I. H. Hillier, *The Calculated Defect Structure of ZnO*, J. Phys. Paris **41** (Suppl. to No. 7), (1980), C6-64.

21. G. Müller, *Optical and Electrical Spectroscopy of Zinc Oxide Crystals Simultaneously Doped with Copper and Donors*, Phys. Status Solidi **(b)76** (1976), 525.

22. R. Pappalardo, D. L. Wood, and R. C. Linares, Jr., *Optical Absorption Study of Co-Doped Oxide Systems: II,* J. Chem. Phys. **35** (1961), 2041.

23. R. Pappalardo, D. L. Wood, and R. C. Linares, Jr., *Optical Absorption Spectra of Ni-Doped Oxide Systems: I,* J. Chem. Phys. **35** (1961), 1460.

24. R. Purlis, A. Jakimavicius, and A. Sirvaitis, *Temperature Dependence of Root-Mean-Square Dynamic Displacement and X-Ray Diffraction Characteristic Temperatures of ZnS and ZnO,* Izv. Vyssh. Uchelbn. Zaved. Fiz. **24** (1981), 115.

25. M. L. Reynolds and G.F.J. Garlick, *The Infrared Emission of Nickel Ion Impurity Centres in Various Solids,* Infrared Phys. **7** (1967), 151.

26. T. M. Sabine and S. Hogg, *The Wurtzite Z Parameter for Beryllium Oxide and Zinc Oxide,* Acta Crystallogr. **B25** (1969), 2254.

27. P. C. Schmidt, A. Weiss, and T. P. Das, *Effect of Crystal Fields and Self-Consistency on Dipole and Quadrupole Polarizabilities of Closed-Shell Ions,* Phys. Rev. **B19** (1979), 5525.

28. H.-J. Schultz, *Zinc Oxide,* in *Current Topics in Materials Science,* K. Kaldis, ed., vol 7, North-Holland, Amsterdam (1981).

29. H.-J. Schultz and M. Thiede, *Optical Spectroscopy of $3d^7$ and $3d^8$ Impurity Configurations in a Wide-Gap Semiconductor (ZnO:Co, Ni, Cu),* Phys. Rev. **B35** (1987), 18.

30. T. Shiraishi, *The Crystal Structure of Zinc Oxide-Chromium (IV) Oxide System Catalysts for Methanol Synthesis,* Niihama Kogyo Koto Semmon Gakko Kiyo **17** (1981), 45.

31. A. E. Tsurkan, L. V. Buzhor, B. I. Kidyarov, and P. G. Pas'kov, *Electrical Properties of Zinc Oxide Single Crystals Prepared by Different Methods,* Poluch. Issled. Nov. Materialov Poluprovodn. Tekhn. (1980), 161.

32. V. P. Vlasov, G. I. Distler, V. M. Kanevskii, and G. D. Shnyrev, *Effects of Impurity Structures on Brittle Fracture in ZnS and ZnO Crystals,* Izv. Akad. Nauk SSSR Ser. Fiz. **44** (1980), 1302.

33. H. A. Weakliem, *Optical Spectra of Ni^{2+}, Co^{2+}, and Cu^{2+} in Tetrahedral Sites in Crystals,* J. Chem. Phys. **36** (1962), 2117.

34. R.W.G. Wyckoff, *Crystal Structures,* vol 1, Interscience, New York (1968), p 111.

35. E. Ziegler, A. Heinrich, H. Oppermann, and G. Stöver, *Electrical Properties and Non-stoichiometry in ZnO Single Crystals,* Phys. Status Solidi **(a)66** (1981), 635.

20. ZnS

20.1 Crystallographic Data on ZnS

20.1.1 Cubic T_d^2 ($F\overline{4}3m$), 216, Z = 4

Ion	Site	Symmetry	x^a	y	z	q	$\alpha\,(\text{Å}^3)^b$
Zn	4a	T_d	0	0	0	2	0.676
S	4c	T_d	1/4	1/4	1/4	−2	4.893

[a]X-ray data: a = 5.4093 Å (Wyckoff, 1968).
[b]Schmidt et al (1979).

20.1.2 Hexagonal C_{6v}^4 ($P6_3mc$), 186, Z = 2

Ion	Site	Symmetry	x	y	z	q	$\alpha\,(\text{Å}^3)$
Zn	2(b)	C_{3v}	1/3	2/3	0	2	0.676
S	2(b)	C_{3v}	1/3	2/3	z	−2	4.893

20.1.3 X-ray data on hexagonal ZnS

a	b	z_s	Ref
3.811	6.234	0.375	57
3.8227	6.2607	0.3748	19

20.2 Crystal Fields

20.2.1 Crystal-field components, A_{nm} (cm^{-1}/Å^n), for Zn (T_d) site of cubic ZnS

$A_{nm}{}^a$	Monopole	Self-induced	Total
A_{32}	15,219	−7349	7869
A_{40}	−4,610	4035	−574.8

[a]$A_{44} = \sqrt{5/14}\ A_{40}$

20.2.2 Crystal-field components, A_{nm} (cm^{-1}/Å^n), for Zn (C_{3v}) site of hexagonal ZnS (Wyckoff, 1968)

A_{nm}	Monopole	Self-induced	Dipole	Total
A_{10}	20,496	0	9,537	30,033
A_{20}	10,188	−3,463	26,174	32,899
A_{30}	18,608	−10,166	7,716	16,158
A_{33}	7,926	−3,754	−4,809	−636.8
A_{40}	3,800	−4,417	3,128	2,511
A_{43}	−3,877	2,887	952.3	−37.21
A_{50}	1,334	−2,505	4,341	3,171
A_{53}	339.5	−518.3	874.1	695.3

20.2.3 Crystal-field components, A_{nm} (cm^{-1}/Å^n), for Zn (C_{3v}) site of hexagonal ZnS (Kisi and Elcombe, 1989)

A_{nm}	Monopole	Self-induced	Dipole	Total
A_{10}	18,323	—	408.0	18,731
A_{20}	559.8	−25.39	18,125	18,659
A_{30}	15,643	−7660	5,382	13,366
A_{33}	9,767	−4908	−4,104	755.1
A_{40}	3183	−2760	−155.7	357.2
A_{43}	−4,017	3221	74.57	−721.0
A_{50}	−80.66	41.89	2,219	2,180
A_{53}	−109.6	−21.32	1,138	1,007

20.3 Experimental Values (cm^{-1}) of $F^{(2)}$, $F^{(4)}$, ζ, and B_{40} for $3d^N$ Ions

Ion	$3d^N$	$F^{(2)}$	$F^{(4)}$	ζ	$B_{40}{}^a$	Ref
V^{2+}	$3d^3$	40,439	28,867	—	−10,500	14
Cr^{2+}	$3d^4$	44,450	35,910	—	−10,710	12
Mn^{2+}	$3d^5$	51,590	39,060	—	−10,500	13
Mn^{2+}	$3d^5$	56,454	32,925	—	−11,603	45
Fe^{2+}	$3d^6$	51,433	36,195	—	−7,486	20,47
Co^{2+}	$3d^7$	49,516	35,438	583	−7,897	56
Co^{2+}	$3d^7$	53,851	40,619	—	−7,509	34
Ni^{2+}	$3d^8$	46,449	32,211	477	−9,321	56
Ni^{2+}	$3d^8$	50,119	29,445	—	−10,701	41

$^a B_{44} = \sqrt{5/14}\ B_{40}$

20.4 Bibliography and References

1. C. Benecke, W. Busse, H.-E. Gumlich, and J.-J. Moros, *Time-Resolved Spectroscopy of the Low-Energy Emission Bands of Highly Doped* $Zn_{1-x}Mn_xS$, Phys. Status Solidi (**b**)142 (1987), 301.

2. S. W. Biernacki, *Parametrization of Crystal Field Spectra of* Mn^{2+} *in ZnSe and ZnS,* Phys. Status Solidi (**b**)132 (1985), 557.

3. S. W. Biernacki and B. Clerjaud, *Optical Spectra of Ti, V, and Co in III-V and II-VI Crystals*, Acta Phys. Pol. **A73** (1988), 251.

4. S. W. Biernacki, G. Roussos, and H.-J. Schulz, *The Luminescence of* V^{2+} *(d^3) and* $V^{3+}(d^2)$ *Ions in ZnS and an Advanced Interpretation of Their Excitation Levels*, J. Phys. **C21** (1988), 5615.

5. S. W. Biernacki, G. Roussos, and H.-J. Schulz, *(Mo-P-23) Excitation Terms of* V^{3+} *and* V^{2+} *Centers in ZnS Crystals*, Acta Phys. Pol. **A73** (1988), 259.

6. D. Boulanger, *Ligand-Field Theory for the Orbit-Lattice and Spin-Lattice Coupling Coefficients of Mn^{2+} in Zns and ZnSe*, J. Cryst. Growth **101** (1990), 368.

7. F. J. Bryant and A. Krier, *Low-Voltage Direct Current Electroluminescence in ZnS:Re Thin Films,* IEE Proc. **130** (1983), 160.

8. D. Curie, C. Barthou, and B. Canny, *Covalent Bonding of Mn^{2+} Ions in Octahedral and Tetrahedral Coordinates,* J. Chem. Phys. **61** (1974), 3048.

9. B. Clerjaud, A. Gelineau, F. Gendron, C. Porte, J. M. Baranowski, and Z. Liro, *Optical Properties of Ni^{2+} (d^9) in ZnS*, Physica **116B** (1983), 500.

10. A. Fazzio, M. J. Caldas, and A. Zunger, *Many-Electron Multiplet Effects in the Spectra of 3d Impurities in Heteropolar Semiconductors,* Phys. Rev. **B30** (1984), 3430.

11. R. Grasser, A. Scharmann, and B. Seidl, *The Influence of High Temperature Annealing on Luminescence and Energy Transfer in ZnS/Mn Crystals*, J. Cryst. Growth **101** (1990), 449.

12. G. Grebe and H.-J. Schultz, *Interpretation of Excitation Spectra of $ZnS:Cr^{2+}$ by Fitting the Eigen Values of the Tanabe-Sugano Matrices,* Phys. Status Solidi **(b)54** (1972), K 69.

13. H. E. Gumlich, R. L. Pfrogner, J. C. Schaffer, and F. E. Williams, *Optical Absorption and Energy Levels of Manganese in ZnS:Mn Crystals,* J. Chem. Phys. **44** (1966), 3929.

14. Le. M. Hoang and J. M. Baranowski, *Absorption and Luminescence of $V(d^3)$ in II-VI Compound Semiconductors,* Phys. Status Solidi **(b)84** (1977), 361.

15. A. Hoffman, R. Heitz, and I. Broser, *Fe^{2+} as Near-Infrared Luminescence Center in ZnS*, Phys. Rev. **B41** (1990), 5806.

16. G. L. House and H. G. Drickamer, *High Pressure Luminescence Studies of Localized Excitations in ZnS Doped with Pb^{2+} and Mn^{2+},* J. Chem. Phys. **67** (1977), 3230.

17. U. G. Kaufmann and P. Koidl, *Jahn-Teller Effect in the $^3T_1(P)$ Absorption Band of Ni^{2+} in Zns and ZnO*, J. Phys. **C7** (1974), 791.

18. U. Kaufmann, P. Koidl, and O. F. Schirmer, *Near Infrared Absorption of Ni^{2+} in ZnO and ZnS: Dynamic Jahn-Teller Effect in the $^3T_2$ State*, J. Phys. **C6** (1973), 310.

19. E. H. Kisi and M. M. Elcombe, *U Parameters for the Wurtzite Structure of Zns and ZnO Using Powder Neutron Diffraction,* Acta Crystallogr. **C45** (1989), 1867.

20. F. F. Kodzhespirov, M. Bulanyi, and J. A. Tereb, *Optical and Thermal Depth of Energy Levels of Photosensitive Paramagnetic Centers,* Sov. Phys. Solid State **16** (1975), 2052.

21. P. Koidl, *Jahn-Teller Effect in the $^4T_1(1)$ and $^4T_2(1)$ States of Tetrahe-drally Coordinated Mn^{2+}*, Phys. Status Solidi (**b**)**74** (1976), 477.

22. P. Koidl and A. Räuber, *Optical Absorption and Electron Spin Resonance of Co^{2+} in Cubic, Hexagonal, and 4H Polytype Zinc Sulfide*, J. Phys. Chem. Solids **35** (1974), 1061.

23. P. Koidl, O. F. Schirmer, and U. Kaufmann, *Near-Infrared Absorption of Co^{2+} in ZnS: Weak Jahn-Teller Coupling in the $^4T_2$ and $^4T_1$ States*, Phys. Rev. **B8** (1973), 4926.

24. B. Lambert, T, Buch, and A. Geoffroy, *Optical Properties of Mn^{2+} in Pure and Faulted Cubic ZnS Single Crystals*, Phys. Rev. **B8** (1973), 863.

25. D. Langer and S. Ibuki, *Zero-Phonon Lines and Phonon Coupling in ZnS:Mn*, Phys. Rev. **A138** (1965), 809.

26. D. Langer and H. Richter, *Zero-Phonon Lines and Phonon Coupling of ZnSe:Mn and CdS:Mn*, Phys. Rev. **146** (1966), 554.

27. H. H. Li, *Refractive Index of ZnS, ZnSe, and ZnTe and Its Wavelength and Temperature Derivatives*, J. Phys. Chem. Ref. Data **13** (1984), 103.

28. W. Low and M. Weger, *Paramagnetic Resonance and Optical Spectra of Divalent Iron in Cubic Fields: I.—Theory*, Phys. Rev. **118** (1960), 1119.

29. W. Low and M. Weger, *Paramagnetic Resonance and Optical Spectra of Divalent Iron in Cubic Fields: II.—Experimental Results*, Phys. Rev. **118** (1960), 1130.

30. S. Mardix, *Polytypism: A Controlled Thermodynamic Phenomena*, Phys. Rev. **B33** (1986), 8677.

31. L. Martinelli, M. Passaro, and G. P. Parravicini, *Multimode Vibronic Model for Fe^{2+} in ZnS*, Phys. Rev. **B40** (1989), 10,443.

32. D. S. McClure, *Optical Spectra of Exchange Coupled Mn^{++} Ion Pairs in ZnS:MnS*, J. Chem. Phys. **39** (1963), 2850.

33. V. Medizadeh and S. Mardix, *New ZnS Polytypes*, Acta Crystallogr. **C42** (1986), 518.

34. J. M. Noras, H. R. Szawelska, and J. W. Allen, *Energy Levels of Cobalt in ZnSe and ZnS*, J. Phys. **C14** (1981), 3255.

35. D. T. Palumbo and J. J. Brown, *Electronic States of Mn^{2+} Activated Phosphors: II.—Orange-to-Red Emitting Phosphors*, J. Electrochem. Soc. **118** (1971), 1159.

36. U. W. Pohl and H.-E. Gumlich, *Long Range Mn Ion Pairs and other Mn Related Centers in ZnS Detected by Site-Selected Spectroscopy*, J. Cryst. Growth **101** (1990), 521.

37. U. W. Pohl, A. Ostermeier, W. Busse, and H. Gumlich, *Influence of Stacking Faults in Polymorphic ZnS on the d^5 Crystal-Field States of Mn^{2+}*, Phys. Rev. **B42** (1990), 5751.

38. M. L. Reynolds and G.F.J. Garlick, *The Infrared Envision of Nickel Ion Impurity Centers in Various Solids,* Infrared Phys. **7** (1967), 151.

39. J. W. Richardson and G. J. M. Janssen, *Theoretical Analysis of Orbital and Correlation Effects on the Electronic Absorption Spectrum of the MnS_4 Center in Zinc-Blende Crystals*, Phys. Rev. **B39** (1989), 4958.

40. G. Roussos, J. Nagel, and H.-J. Schultz, *Luminescent Ni^{2+} Centers and Changes of the Charge State of Nickel Ions in ZnS and ZnSe*, Z. Phys. **B53** (1983), 96.

41. G. Roussos and H.-J. Schultz, *A New Infrared Luminescence of Nickel-Doped ZnS and Its Interpretation by Means of Absorption Spectroscopy with Ni^{2+} Ions,* Phys. Status Solidi **(b)100** (1980), 577.

42. P. C. Schmidt, A. Weiss, and T. P. Das, *Effect of Crystal Fields and Self-Consistency on Dipole and Quadrupole Polarizabilities of Closed-Shell Ions,* Phys. Rev. **B19** (1979), 5525.

43. J. Schneider and A. Räuber, *Electron Spin Resonance of Ti^{2+} in ZnS,* Phys. Lett. **21** (1966), 380.

44. H.-J. Schultz, G. Roussos, and S. W. Biernacki, *Optical Properties of Vanadium Ions in Zinc Sulphide*, Z. Naturforsch. **45a** (1990), 669.

45. H.-J. Schultz and M. Thiede, *Optical Spectroscopy of $3d^7$ and $3d^8$ Impurity Configurations in a Wide-Gap Semiconductor (ZnO: Co, Ni, Cu),* Phys. Rev. **B35** (1987), 18.

46. M. Skowronski, M. Godlewski, and Z. Liro, *The Fe^{2+} Ion Energy Scheme in ZnS Crystals*, Proc. Conf. Phys. **2** (1981), 112.

47. M. Skowronski and Z. Liro, *Spin-Forbidden Transitions in the Fe^{2+} Ion in ZnS Crystals,* J. Phys. **C15** (1982), 137.

48. G. A. Slack, F. S. Ham, and R. M. Chrenko, *Optical Absorption of Tetrahedral Fe^{2+} $(3d^6)$ in Cubic ZnS, CdTe, and $MgAl_2O_4$*, Phys. Rev. **152** (1966), 376–402.

49. G. A. Slack and B. M. O'Meara, *Infrared Luminescence of Fe^{2+} in ZnS,* Phys. Rev. **163** (1967), 335.

50. G. A. Slack, S. Roberts, and F. S. Ham, *Far-Infrared Optical Absorption of Fe^{2+} in ZnS*, Phys. Rev. **155** (1967), 170.

51. V. I. Sokolov, A. N. Mamedov, T. P. Surkova, M. V. Chukichev, and M. P. Kulakov, *Energy States of Cobalt in Zinc Selenide and Zinc Sulphide*, Opt. Spectrosc. **62** (1987), 480.

52. J. T. Vallin, G. A. Slack, S. Roberts, and A. E. Hughes, *Infrared Absorption in Some II-VI Compounds Doped with Cr*, Phys. Rev. **B2** (1970), 4313.

53. J. T. Vallin and G. D. Watkins, *EPR of Cr^{2+} in II-VI Lattices*, Phys. Rev. **B9** (1974), 2051.

54. V. P. Vlasov, G. I. Distler, V. M. Kanevskii, and G. D. Shnyrev, *Effects of Impurity Structures on Brittle Fracture in ZnS and ZnO Crystals*, Izv. Akad. Nauk SSSR Ser. Fiz. **44** (1980), 1302.

55. D. Wasik, M. Baj, and Z. Luro, *Pressure-Dependent Coupling of the $^3A_2(F)$ and $^1T_2(D)$ States of Ni^{2+} in ZnS and ZnSe*, Acta Phys. Pol. **A79** (1991), 319.

56. H. A. Weakliem, *Optical Spectra of Ni^{2+}, Co^{2+} and Cu^{2+} in Tetrahedral Sites in Crystals*, J. Chem. Phys. **36** (1962), 2117.

57. R.W.G. Wyckoff, *Crystal Structures*, vol 1, Interscience, New York (1968).

58. Wan-Lun Yu, *Crystal Field Effect on the g-Factors of ^6S-State Ions*, Phys. Status Solidi **(b)158** (1990), K13.

59. Zhao Sang-Bo, Wang Hui-Su, and Wu Ping-Feng, *An EPR Theory for C_{2v} (d^6) and the Spectral Analysis of $ZnS:Fe^{2+}$*, J. Phys. Cond. Mater. **2** (1990), 687.

60. Sang-Bo Zhao, Hui-Su Wang, and Jun-Kai Xie, *Coordination Symmetry and Energy Spectrum Analysis of Fe^{2+} in $ZnS:Fe^{2+}$ Crystals*, Phys. Status Solidi **(b)151** (1989), 167.

61. M. Zigone, R. Beserman, and B. Lambert, *The Effect of Crystalline Structure on Zero-Phonon and Phonon-Assisted Emission in ZnS:Mn*, J. Lumin. **9** (1974), 45.

21. K₂PtCl₆

21.1 Crystallographic Data on K₂PtCl₆

21.1.1 Cubic O_5^h (Fm3m), 225, Z = 4

Ion	Site	Symmetry	x	y	z	q	α (Å³)
Pt	4(a)	O_h	0	0	0	4	0.67[a]
K	8(c)	T_d	1/4	1/4	1/4	1	0.827[b]
Cl	24(e)	C_{4v}	x	0	0	−1	2.694[b]

[a]Fraga et al (1976).
[b]Schmidt et al (1979).

21.1.2 X-ray data on K₂PtCl₆

a (Å)	X_{Cl}	Ref
9.755	0.240	18
9.6911	0.23839	17

21.2 Crystal-Field Components, A_{nm} (cm⁻¹/Åⁿ), for Pt (O_h) Site

A_{nm}[a]	Monopole	Self-induced	Dipole	Total	Ref
A_{40}	6115	−5017	10,480	11,578	18
A_{40}	6453	−5480	11,286	12,258	17

[a]$A_{44} = \sqrt{5/14}\ A_{40}$

21.3 Experimental Parameters (cm⁻¹)

Ion	nd^N	$F^{(2)}$	$F^{(4)}$	ζ	B_{40}[a]	Ref
Ru⁴⁺	$4d^4$	48,888	17,174	1044	39,732	13
Re⁴⁺	$5d^3$	28,749	22,907	2392	63,729	5
Os⁴⁺	$5d^4$	45,381	16,330	2416	47,229	4
Re⁴⁺	$5d^3$	28,843	22,582	2360	63,477	7
Re⁴⁺	$5d^3$	29,963	22,907	2392	63,729	9

[a]$B_{44} = \sqrt{5/14}\ B_{40}$

21.4   Bibliography and References

1. A. M. Black and C. D. Flint, *Luminescence Spectra and Relaxation Processes of $ReCl_6^{2-}$ in Cubic Crystals*, J. Chem. Soc. Faraday Trans. 2 **73** (1976), 877.

2. P. B. Dorain, *Magnetic and Optical Properties of Transition Metal Ions in Single Crystals,* Aerospace Laboratories Report, ARL-73-0139 (October 1973), NTIS AD 769870.

3. P. B. Dorain, *The Spectra of Re^{4+} in Cubic Crystal Fields,* in *Transition Metal Chemistry,* R. L. Carlin, ed., vol 4, Marcel Dekker, New York (1968), pp 1–31.

4. P. B. Dorain, H. H. Patterson, and P. C. Jordan, *Optical Spectra of Os^{4+} in Single Cubic Crystals at $4.2°K$,* J. Chem. Phys. **49** (1968), 3845.

5. P. B. Dorain and R. G. Wheeler, *Optical Spectrum of Re^{4+} in Single Crystals of K_2PtCl_6 and Cs_2ZrCl_6 at $4.2°K$,* J. Chem. Phys. **45** (1966), 1172.

6. S. Fraga, K.M.S. Saxena, and J. Karwowski, *Handbook of Atomic Data,* Elsevier, New York (1976).

7. C. D. Flint and A. G. Paulusz, *High Resolution Infrared and Visible Luminescence Spectra of $ReCl_6^{2-}$ and $ReBr_6^{2-}$ in Cubic Crystals,* Mol. Phys. **43** (1981), 321.

8. J. Gilchrist, *Low Temperature Dielectric Relaxation in Ammonium Hexachlorostannate and Some Other Antifluorites,* J. Phys. Chem. Solids **50** (1989), 857.

9. Han Yande, Li Baifu, Zhu Yukui, and Sun Chia-Chung, *The Analysis of the Ligand Field Theory of Crystal Spectrum of d^3 (Re^{4+} and d^4 (Os^{+4}) Configuration,* Chem. J. Chinese Univ. **3** (1982), 97.

10. S. M. Khan, H. H. Patterson, and H. Engstrom, *Multiple State Luminescence for the d^4 $OsCl_6^{2-}$ Impurity Ion in K_2PtCl_6 and Cs_2ZrCl_6 Cubic Crystals,* Mol. Phys. **35** (1978), 1623.

11. Li Xuequi and Cheng Yu, *A Program for the Calculation of Ligand Field Theory,* Chem. J. Chinese Univ. **5** (1984), 725.

12. S. Maniv, J. Bronstein, and W. Low, *Electron-Paramagnetic-Resonance Spectrum of Tc^{4+} in Single Crystals of K_2PtCl_6,* Phys. Rev. **187** (1969), 403.

13. H. Patterson and P. Dorain, *Optical Spectra of Ru^{4+} in Single Crystals of K_2PtCl_6 and Cs_2ZrCl_6 at $4.2°K$,* J. Chem. Phys. **52** (1970), 849.

14. P. C. Schmidt, A. Weiss, and T. P. Das, *Effect of Crystal Fields and Self-Consistency on Dipole and Quadrupole Polarizabilities of Closed-Shell Ions,* Phys. Rev. **B19** (1979), 5525.

15. H.-H. Schmidtke and D. Strand, *The Emission Spectrums of OsCl$_6^{2-}$ Doped in Various Cubic Host Lattices,* Inorg. Chim. Acta **62** (1982), 153.

16. N. Schoenen and H.-H. Schmidtke, *Linear Polarized Absoprtion Spectra ReX$_6^{2-}$ in Trigonal Crystals*, Chem. Phys. Lett. **95** (1983), 497.

17. H. Takazawa, S. Ohba, and Y. Saito, *Electron-Density Distribution in Crystals of K$_2$[MCl$_6$] (M = Re, Os, Pt) and K$_2$[PtCl$_4$] at 120K,* Acta Crystallogr. **B46** (1990), 166.

18. R.W.G. Wyckoff, *Crystal Structures,* vol 3, Interscience, New York (1968), p 339.

19. R. K. Yoo, B. A. Kozikowski, S. C. Lee, and T. A. Keiderling, *Visible Region Absorption and Excitation Spectroscopy of K$_2$ReCl$_6$ and Various ReCl$_6^{2-}$ Containing A$_2$MCl$_6$ Host Crystals*, Chem. Phys. **117** (1987), 255.

20. R. K. Yoo, S. C. Lee, B. A. Kozikowski, and T. A. Keiderling, *Intraconfigurational Absorption Spectroscopy of ReCl$_6^{2-}$ in Various A$_2$MCl$_6$ Host Crystals*, Chem. Phys. **117** (1987), 237.

22. $Y_3Ga_5O_{12}$ (YGG)

22.1 Crystallographic Data on $Y_3Ga_5O_{12}$

Cubic O (Ia3d), 230, Z = 8

Ion	Site	Symmetry	x^a	y	z	q	$\alpha\,(\text{Å}^3)^b$
Y	24(c)	D_2	0	1/4	1/8	3	0.870
Ga_1	16(a)	C_{3i}	0	0	0	3	0.458
Ga_2	24(d)	S_4	0	1/4	3/8	3	0.458
0	96(h)	C_1	−0.0272	0.05580	0.1501	−2	1.349

^{a}X-ray data: a = 12.28 (Euler and Bruce, 1965).
^{b}Schmidt et al (1979).

22.2 Crystal-Field Components, A_{nm} (cm^{-1}/Å^n)

22.2.1 For Ga_2 ion in 24(d) (S4) site

A_{nm}	Monopole	Self-induced	Dipole	Total
A_{20}	7,529	−2139	9280	14,670
ReA_{32}	−19,875	5487	−9441	−23,828
ImA_{32}	32,745	−8849	5911	29,807
A_{40}	−19,871	8062	−6494	−18,303
ReA_{44}	−3,959	1499	1012	−1,448
ImA_{44}	−6,411	2935	−2624	−6,100
ReA_{52}	−2,136	1449	−2205	−2,892
ImA_{52}	3,719	−2389	2509	3,839
$|A_{44}|$	7,535	—	—	6,270

22.2.2 For Ga_1 ion in 16(a) (C_{3i}) site (rotated so that z-axis is parallel to (111) crystallographic axis)

A_{nm}	Monopole	Self-induced	Dipole	Total
A_{20}	15,183	−1576	−14,849	−1,242
A_{40}	−17,784	6625	1,754	−9,403
ReA_{43}	1,295	−745.1	4,956	5,505
ImA_{43}	−18,678	6651	530.7	−11,496
$|A_{44}|$	18,722	—	—	12,746

22.3 Experimental Parameters (cm⁻¹)

Ion	$F^{(2)}$	$F^{(4)}$	B_{20}	B_{40}	B_{43}	Ref
Cr^{3+}	53,994	40,735	—	$-22,820^a$	—	17
Cr^{3+}	54,638	40,456	—	$-21,140^a$	—	17
Cr^{3+}	52,920	37,926	—	$-23,100^a$	—	6
Cr^{3+}	—	—	—	$-21,441^a$	—	15
Cr^{3+}	51,730	42,840	—	$-23,100^a$	—	11
Co^{2+}	56,350	39,690	—	$-8,190^b$	—	13

^{a}Cubic approximation $B_{43} = \sqrt{10/7}\ B_{40}$

^{b}Cubic (tetrahedral) approximation $B_{44} = \sqrt{5/14}\ B_{40}$.

22.4 Bibliography and References

1. M. Kh. Ashurov, Yu. K. Voronko, V. V. Osiko, A. A. Sobol, B. P. Starikov, M. I. Timoshechkin, and A. Ya. Yablonskii, *Inequivalent Luminescence Centers of Er^{3+} in Gallium Garnet Single Crystals,* Phys. Status Solidi (a)**35** (1976), 645.

2. G. Burns, E. A. Geiss, B. A. Jenkins, and M. I. Nathan, *Cr^{3+} Fluorescence in Garnets and Other Crystals,* Phys. Rev. **139** (1965), A1687.

3. J. W. Carson and R. L. White, *Zero-Field Splitting of the Cr^{3+} Ground State in YGa and YAl Garnet,* J. Appl. Phys. **32** (1961), 1787.

4. J. R. Chamberlain and R. W. Cooper, *Paramagnetic Resonance in Yttrium Gallium Garnet: Co^{2+} and Mn^{2+},* Proc. Phys. Soc. **87** (1966), 967.

5. F. Euler and J. A. Bruce, *Oxygen Coordinates of Compounds with Garnet Structure,* Acta Crystallogr. **19** (1965), 971.

6. J. Ferguson and D. L. Wood, *Crystal Field Spectra of $d^{3,7}$ Ions: VI. The Weak Field Formalism and Covalency,* Aust. J. Chem. **23** (1970), 861.

7. S. Geschwind, *Paramagnetic Resonance of Fe^{3+} in Octahedral and Tetrahedral Sites in Yttrium Gallium Garnet (YGaG) and Anisotropy of Yttrium Iron Garnet (YIG),* Phys. Rev. **121** (1961), 363.

8. S. Haussühl and W. Effgen, *Faraday Effect in Cubic Crystals,* Z. Kristallogr. **183** (1988), 153.

9. B. Henderson, A. Marshall, M. Yamaga, K. P. O'Donnell, and B. Cockayne, *The Temperature Dependence of Cr^{3+} Photoluminescence in Some Garnet Crystals,* J. Phys. **C21** (1988), 6187.

10. B. Henderson, K. P. O'Donnell, M. Yamaga, B. Cockayne, and M. J. P. Payne, *High-Resolution Spectroscopy of Cr^{3+} Ions in Solids,* AIP Conf. Proc. **172** (1988), 425.

11. A. D. Liehr, *The Three Electron (or Hole) Cubic Ligand Field Spectrum*, J. Phys. Chem. **67** (1963), 1314.

12. A. Marshall, K. P. O'Donnell, M. Yamaga, B. Henderson, and B. Cockayne, *Disorder and the Shape of the R-Lines in Cr^{3+}-Doped Garnets*, Appl. Phys. **A50** (1990), 565.

13. R. Pappalardo, D. L. Wood, and R. C. Linares, Jr., *Optical Absorption Study of Co-Doped Oxide Systems: II,* J. Chem. Phys. **35** (1961), 2041.

14. R. Pappalardo, D. L. Wood, and R. C. Linares, Jr., *Optical Absorption Spectra of Ni-Doped Oxide Systems: I, J.* Chem. Phys. **35** (1961), 1460.

15. Yuanwu Qiu, Xiao Tang, and Ji-kang Zhu, *X_α Studies of Laser Crystals II. The Effect of the Crystal Field Strength on the Electronic Structures of Cr^{3+} in Garnets*, Z. Phys. **B78** (1990), 271.

16. P. C. Schmidt, A. Weiss, and T. P. Das, *Effect of Crystal Fields and Self-Consistency on Dipole and Quadrupole Polarizabilities of Closed-Shell Ions,* Phys. Rev. **B19** (1979), 5525.

17. B. Struve and G. Huber, *The Effect of the Crystal Field Strength on the Optical Spectra of Cr^{3+} in Gallium Garnet Laser Crystals,* Appl. Phys. **B36** (1985), 195.

18. M. D. Sturge, F. R. Merritt, J. C. Hensel, and J. P. Remeika, *Magnetic Behavior of Cobalt in Garnets: I. Spin Resonance in Cobalt-Doped Yttrium Gallium Garnet,* Phys. Rev. **180** (1969), 402.

19. D. L. Wood, J. Ferguson, K. Knox, and J. F. Dillon, Jr., *Crystal-Field Spectra of $d^{3,7}$ Ions: III.—Spectrum of Cr^{3+} in Various Octahedral Crystal Fields,* J. Chem. Phys. **39** (1967), 3595.

20. D. L. Wood and J. P. Remeika, *Optical Absorption of Tetrahedral Co^{3+} and Co^{2+} in Garnets,* J. Chem. Phys. **46** (1967), 3595.

21. M. Yamaga, B. Henderson, K. P. O'Donnell, C. T. Cowan, and A. Marshall, *Temperature Dependence of the Lifetime of Cr^{3+} Luminescence in Garnet Crystals I*, Appl. Phys. **B50** (1990), 425.

22. M. Yamaga, B. Henderson, K. P. O'Donnell, and G. Yue, *Temperature Dependence of the Lifetime of Cr^{3+} Luminescence in Garnet Crystals II. The Case of YGG*, Appl. Phys. **B51** (1990), 132.

23. Min Yin, Li-Ren Lou, and Shao-Hong Xu, *New Luminescent Centre in YGG:Cr Crystal*, J. Lumin. **40,41** (1988), 631.

23. $La_{2-x}Sr_xCuO_4$

23.1 Crystallographic Data on $La_{1.85}Sr_{0.15}CuO_4$

23.1.1 Tetragonal, D_{4h}^{17} ($I4/mmm$), 139, $Z = 2$; $T = 300$ K

Ion	Site	Symmetry	x	y	z	q^a	q^b
La/Sr	4(e)	C_{4v}	0	0	z	2.775	2.925
Cu	2(a)	D_{4h}	0	0	0	2.450	2.15
O_1	4(e)	C_{4v}	0	0	z	-2	-2
O_2	4(c)	D_{2h}	0	1/2	0	-2	-2

[a]The effective charges on the Cu ions are chosen as 55% Cl^{2+} + 45% Cu^{3+}, and the total charge in the unit cell vanishes.

[b]The charge on La is taken as 3 and the charge on Sr is taken as 2, so that the average charge on the La/Sr site is $[3(1.85) + 2(0.15)]/2$. The charge on the Cu ion is then chosen so that the total charge in a unit cell is zero.

23.1.2 Orthorhombic D_{2h}^{18} ($Cmca$), 64, $Z = 4$; $T = 10$ K and 60 K

Ion	Site	Symmetry	x	y	z	q^a	q^b
La/Sr	8(f)	C_s	0	y	z	2.775	2.925
Cu	4(a)	C_{2h}	0	0	0	2.450	2.15
O_1	8(f)	C_s	0	y	z	-2	-2
O_2	8(e)	C_2	1/4	y	1/4	-2	-2

[a,b]See notes to 23.1.1

23.2 X-Ray Data

23.2.1 Tetragonal, $T = 300$ K

a	c	Z_{La}	Z_{O1}	Ref
3.7793	13.226	0.36046	0.1824	3
3.7749	13.2231	0.3606	0.1826	6

23.2.2 Orthorhombic, $T = 10$ K and 60 K

Temp	a	b	c	y_{La}	z_{La}	y_{O1}	z_{O1}	y_{O2}
10 K	5.3240	13.1832	5.3547	-0.36077	-0.00496	-0.18260	0.0255	-0.00573
60 K	5.3252	13.1844	5.3546	-0.36072	-0.00495	-0.18257	0.0256	-0.00560

Cava et al (1987) (transformed to the standard in the International Tables).

23.3 Crystal-Field Components, A_{nm} (cm^{-1}/Å^n)

23.3.1 For La (C_{4v}) site in tetragonal $La_{1.85}Sr_{0.15}CuO_4$

A_{nm}	300 K[a]	300 K
A_{10}	−6369	4915
A_{20}	8888	924
A_{30}	−4634	−5675
A_{40}	1954	1451
A_{44}	−1773	−1818
A_{50}	−3188	−3127
A_{54}	1691	1590
A_{60}	364.2	358.4
A_{64}	745.9	720.3
A_{70}	136.4	−123.5
A_{74}	76.61	77.69

[a]Cava et al (1987).

[b]Geiser et al (1987).

23.3.2 For La (C_s) site of orthorhombic $La_{1.85}Sr_{0.15}CuO_4$ rotated so that z-axis of A_{nm} is parallel to b-axis and A_{22} is real and positive

A_{nm}	60 K	10 K	A_{nm}	60 K	10 K
ReA_{11}	3904	3919	ReA_{55}	1657	1711
ImA_{11}	6897	6952	ImA_{55}	1010	1084
A_{20}	−4544	−4534	A_{60}	404.8	405.6
A_{22}	5272	5284	ReA_{62}	−46.89	−47.26
ReA_{31}	−2027	−2023	ImA_{62}	81.61	82.61
ImA_{31}	−30.36	−26.08	ReA_{64}	−688.5	−685.5
ReA_{33}	2759	2760	ImA_{64}	−305.5	−314.3
ImA_{33}	−187.2	−149.7	ReA_{66}	−22.47	−22.21
A_{40}	2573	2567	ImA_{66}	−32.74	−32.38
ReA_{42}	496.0	499.0	ReA_{71}	−83.33	−83.31
ImA_{42}	−184.5	−189.9	ImA_{71}	−2.75	−2.87
ReA_{44}	1444	1439	ReA_{73}	38.85	39.08
ImA_{44}	−260.8	−245.5	ImA_{73}	−27.77	−27.61
ReA_{51}	2070	1992	ReA_{75}	9.35	9.67
ImA_{51}	366.8	359.0	ImA_{75}	20.60	20.67
ReA_{53}	48.29	−104.4	ReA_{77}	85.19	84.98
ImA_{53}	466.5	407.40	ImA_{77}	13.54	15.27

23.4 Bibliography and References

1. P. Boni, J. D. Axe, G. Shirane, R. J. Birgeneau, D. R. Gabbe, H. P. Jenssen, M. A. Kastner, P. J. Picone, T. R. Thurston, M. Sato, and S.

Shamoto, *Lattice Instability in Single-Crystal $La_{2-x}Sr_xCuO_4$*, Physica **B156,157** (1989), 902.

2. G. Burns and F. H. Dacol, *Anomalous Raman Spectra from La_2CuO_4*, Phys. Rev. **B41** (1990), 4747.

3. R. J. Cava, A. Santoro, D. W. Johnson, and W. W. Rhodes, *Crystal Structure of the High-Temperature Superconductor $La_{1.85}Sr_{0.15}CuO_4$ Above and Below T_c*, Phys. Rev. **B35** (1987), 6716.

4. R. T. Collins, Z. Schlesinger, G. V. Chandrashekhar, and M. W. Shafer, *Infrared Study of Anisotropy in Single-Crystal $La_{2-x}Sr_xCuO_4$*, Phys. Rev. **B39** (1989), 2251.

5. V. Geiser, M. A. Bens, A. J. Schultz, H. H. Wang, T. J. Allen, M. R. Monaghan, and J. M. Williams, *Structure Instability in Single Crystals of the High-T_c Superconductor $La_{2-x}Sr_xCuO_4$*, Phys. Rev. **B35** (1987), 6721.

6. T. Kajitani, T. Onozuka, Y. Yamaguchi, M. Hirabayeshi, and Y. Syono, *Displacement Waves in La_2CuO_{4-w} and $La_{1.85}Sr_{0.15}CuO_{4-w}$*, Jpn. J. Appl. Phys. **26** (1987), L1877.

7. K. Ohbayashi, H. Tukamoto, H. Yamasita, A. Fukumoto, Y. Utsunomiya, N. Ogita, M. Udagawa, and S. Funahashi, *Correlation of Superconducting with the Infrared Anomaly in $(La_{1-x}Mx)_2(Cu_{1-y}Ny)O_4$ (M = Ca, Sr and Ba; N = Ni and Zn)*, J. Phys. Soc. Jpn. **59** (1990), 1372.

8. V. I. Simonov, L. A. Muradyan, R. A. Tamazyan, V. V. Osiko, V. M. Tatarinsev, and K. Gamayumov, *Distribution of Sr Atoms in Single Crystals of $(La_{1-x}Sr_x)_2 CuO_{4-\delta}$ and the Superconducting Transition Temperature*, Physica **C169** (1990), 123.

9. B. M. Tissue, K. M. Cirillo, J. C. Wright, M. Daeumling, and D. C. Larbalestier, *Conversion of Lanthanum Copper Oxide ($La_2CuO_{4-\delta}$) to a Superconductor by Treatment in Fluorine Gas*, Solid State Commun. **65** (1988), 51.

10. B. M. Tissue and J. C. Wright, *Identification of Fluorescing Phases in Eu^{3+} Doped Lanthanum Strontium Copper Oxide ($La_{1.85}Sr_{0.15}CuO_4$) Superconductors*, J. Lumin. **42** (1988), 173.

11. B. M. Tissue and J. C. Wright, *Laser Spectroscopy as a Probe of Structural Changes in Europium Doped Lanthanum Strontium Cuprate ($La_{2-x}Sr_xCuO_4$)*, J. Lumin. **40,41** (1988), 313.

12. B. M. Tissue and J. C. Wright, *Observation of Sharp-Line Lanthanide Fluorescence in High Temperature Superconductors*, J. Lumin. **37** (1987), 117.

13. N. Wada, H. Muro-Oka, Y. Nakamura, and Ken-ici Kumagi, *Observation of Static Magnetic Moment on the Copper Site in Superconducting $La_{2-x}Sr_xCuO_4$: The Nuclear Heat Capacity Measurement*, Physica **C157** (1989), 453.

24. Al_2O_3 (Corundum)

24.1 Crystallographic Data on Al_2O_3

Hexagonal D_{3d}^6 ($R3c$) (second setting), 167, $Z = 6$

Ion	Site	Symmetry	x	y	z	q	$\alpha\,(\text{Å}^3)$
Al	12(c)	C_3	0	0	z	3	0.053
O	18(e)	C_2	x	0	1/4	−2	1.349

24.2 X-Ray Data

a	c	z_{Al}	x_O	Ref	Set
4.7628	13.0032	0.352	0.306	56	1
4.7586	12.9897	0.3518	0.6918	73	2
4.75855	12.9906	0.35200	0.6936	73	3
4.75999	12.99481	0.35219	0.69367	73	4
4.7640	13.0091	0.35221	0.30636	17	5

24.3 Crystal-Field Components, A_{nm} $(\text{cm}^{-1}/\text{Å}^n)$, for Al (C_3) Site

Set 1

A_{nm}	Monopole	Self-induced	Dipole	Total
A_{10}	6,472	—	−6424	48.41
A_{20}	−4,896	778.2	1933	−2,185
A_{30}	−9,718	4215	620.4	−4,883
ReA_{33}	−639.7	−1220	1126	−733.6
ImA_{33}	−10,955	3434	1253	−6,268
A_{40}	−18,418	6479	−66.36	−12,005
ReA_{43}	4,371	−1322	629.3	3,678
ImA_{43}	−23,006	9736	417.6	−12,853
A_{50}	8,476	−4130	458.8	4,804
ReA_{53}	1,581	−486.9	107.8	1,201
ImA_{53}	1,766	−592.6	798.9	1,973

Set 2

A_{nm}	Monopole	Self-induced	Dipole	Total
A_{10}	8,456	—	−6841	615
A_{20}	−5,475	852.5	2003	−2,619
A_{30}	−9,692	4327	695.7	−4,669
ReA_{33}	−1,050	−1147	1241	−955.1
ImA_{33}	11,217	−3561	−1337	6,319
A_{40}	−18,612	6555	−65.65	−12,123
ReA_{43}	4,167	−1252	706.9	3,622
ImA_{43}	23,259	−9890	−440.6	12,928
A_{50}	8,352	−4057	487.9	4,783
ReA_{53}	1,564	−470.9	128.8	1,222
ImA_{53}	−1,543	413.5	−861.5	−1,991

Set 3

A_{nm}	Monopole	Self-induced	Dipole	Total
A_{10}	6,840	—	−6447	393.1
A_{20}	−5,026	799.8	1933	−2,293
A_{30}	−9,783	4267	629.7	−4,886
ReA_{33}	−704.1	−1214	1140	−778.5
ImA_{33}	11,064	−3485	−1261	6,318
A_{40}	−18,498	6524	-64.96	−12,039
ReA_{43}	4,355	−1319	639.6	3,676
ImA_{43}	23,136	−9822	−420.9	12,893
A_{50}	8,497	−4150	460.7	4,808
ReA_{53}	1,583	−487.1	110.6	1,206
ImA_{53}	−1,733	561.8	−805.5	−1,976

Set 4

A_{nm}	Monopole	Self-induced	Dipole	Total
A_{10}	6,761	—	−5943	818.8
A_{20}	−5,035	816.1	1791	−2,428
A_{30}	−9,998	4308	577.8	−5,112
ReA_{33}	−634.8	−1221	1055	−800.7
ImA_{33}	11,189	−3526	−1166	6,498
A_{40}	−18,347	6471	−54.40	−11,931
ReA_{43}	4,379	−1323	590.2	3,646
ImA_{43}	23,079	−9795	−395.4	12,889
A_{50}	8,516	−4167	419.4	4,768
ReA_{53}	1,517	−485.0	100.7	1,187
ImA_{53}	−1,728	551.5	−738.8	−1,915

Set 5

A_{nm}	Monopole	Self-induced	Dipole	Total
A_{10}	6,778	—	−582	957.6
A_{20}	−5,003	813.0	1753	−2,437
A_{30}	−10,007	4292	564.1	−5,151
ReA_{33}	−634.0	−1213	1033	−813.9
ImA_{33}	−11,173	3512	1140	−6,521
A_{40}	−18,251	6420	−51.53	−11,882
ReA_{43}	4,361	−1314	577.4	3,625
ImA_{43}	−22,971	9724	387.6	−12,860
A_{50}	8,469	−4134	408.8	4,744
ReA_{53}	1,563	−481.0	98.52	1,180
ImA_{53}	1,709	−539.5	720.8	1,891

24.4 Experimental Parameters (cm^{-1}) for Transition-Metal Ions in Al_2O_3

Ion	nd^N	$F^{(2)}$	$F^{(4)}$	α	ζ	B_{20}	B_{40}	B_{43}	Ref
Ti^{3+}	$3d^1$	—	—	0	120	−997	−23,970	−32,893	42
Ti^{3+}	$3d^1$	—	—	0	—	100	−25,337	−32,435	48
Ti^{3+}	$3d^1$	—	—	0	—	—	$−26,670^a$	—	26
V^{3+}	$3d^2$	—	—	0	—	960	−23,220	−29,819	48
V^{3+}	$3d^2$	47,390	31,500	0	155	234	−23,564	30,803	40
V^{3+}	$3d^2$	44,660	32,760	0	—	1,200	−22,900	−29,952	58
V^{3+}	$3d^2$	48,720	35,658	0	164	−2,303	−21,165	−30,669	59
V^{2+}	$3d^3$	47,625	25,455	79	—	—	$−21,350^a$	—	67
V^{2+}	$3d^3$	44,458	30,239	0	—	—	$−22,036^a$	—	71
Cr^{3+}	$3d^3$	53,690	39,312	0	180	−1123	−22,350	31,538	41
Cr^{3+}	$3d^3$	52,605	37,800	70	—	−1123	−22,400	−31,622	20
Cr^{3+}	$3d^3$	54,748	36,369	0	—	—	$−25,340^a$	—	71
Cr^{3+}	$3d^3$	54,460	41,580	0	172	—	$−25,200^a$	—	34
Mn^{4+}	$3d^3$	—	—	0	—	—	$−29,820^a$	—	71
Cr^{3+}	$3d^3$	—	—	0	—	1425	−23,510	−31,166	48
Mn^{3+}	$3d^4$	—	—	0	—	1950	−24,658	−33,667	48
Fe^{3+}	$3d^5$	—	—	0	—	—	$−23,100^a$	—	48
Co^{3+}	$3d^6$	—	—	0	—	720	−24,660	−31,023	48
Cu^{3+}	$3d^8$	—	—	0	428	—	−29,400	—	10

$^a B_{43} = \sqrt{10/7}\, B_{40}$

24.5 Bibliography and References

1. R. L. Aggarwal, A. Sanchez, M. M. Stuppi, R. E. Fahey, A. J. Strauss, W. R. Rapoport, and C. P. Khattak, *Residual Infrared Absorption in As-Grown and Annealed Crystals of Ti:Al_2O_3*, IEEE J. Quantum Electron. **QE-24** (1988), 1003.

2. J. O. Artman and J. C. Murphy, *Lattice Sum Evaluations of Ruby Spectral Parameters*, Bull. Am. Phys. Soc. **8** (1963), 323. Also, Phys. Rev. **135** (1964), 1622.

3. C. H. Bair, P. Brockman, R. V. Hess, and E. A. Modlin, *Demonstration of Frequency Control and CW Diode Laser Injection Control of a Titanium-Doped Sapphire Ring Laser with No Internal Optical Elements*, IEEE J. Quantum Electron. **QE-24** (1988), 1045.

4. P. Ballmer, H. Blum, W. J. Borer, K. Eigenmann, and Hs. H. Günthard, *Solid State Reactions and Defects in Verneuil Laser Rubies II*, Helv. Phys. Acta **43** (1970), 829.

5. A. S. Barker, *Infrared Lattice Vibrations and Dielectric Dispersion in Corundum*, Phys. Rev. **132** (1963), 1474.

6. J. C. Barnes, N. P. Barnes, and G. E. Miller, *Master Oscillator Power Amplifier Performance of Ti:Al_2O_3*, IEEE J. Quantum Electron. **QE-24** (1988), 1029.

7. N. P. Barnes, J. A. Williams, J. C. Barnes, and G. E. Lockard, *A Self-Injection Locked, Q-Switched, Line-Narrowed Ti:Al_2O_3 Laser*, IEEE J. Quantum Electron. **QE-24** (1988), 1021.

8. S. A. Basun, A. A. Kaplyanskii, V. K. Sevast'yanov, L. S. Starostina, S. P. Feofilov, and A. A. Chernyshev, *Optical Investigations of Al_2O_3:Ti^{3+} Crystals in an Electric Field: Detection of Two-Stage Photo-Ionization of Ti^{3+} Atoms and of the Linear Stark Effect in Their Spectra*, Sov. Phys. Solid State **32** (1990), 1109.

9. S. A. Batishche, A. A. Demidovich, V. G. Koptev, V. P. Mikhailov, V. A. Mostorvnikov, V. P. Orekhova, B. K. Sevast'yanov, G. A. Skripko, L. S. Starostina, G. A. Tatura, A. P. Shkadarevich, K. F. Chirkina, and A. P. Chirkin, *Generation of Subnanosecond Pulses in Ti^{3+}:Al_2O_3 Crystals*, Sov. Phys. Dokl. **34** (1989), 231.

10. W. E. Blumberg, J. Eisinger, and S. Geschwind, *Cu^{3+} Ion in Corundum*, Phys. Rev. **130** (1963), 900.

11. O. N. Boksha, T. M. Varina, and A. A. Popova, *Optical Spectra of Cr and Mn in Synthetic MgO-Al_2O_3 Spinels*, Sov. Phys. Crystallogr. **17** (1973), 940.

12. O. N. Boksha, T. M. Varina, A. A. Popova, and E. F. Smirnova, *Conditions for the Synthesis and the Optical Spectra of Crystals Containing*

Transition Elements: II.—Corundum Containing Ti, Sov. Phys. Crystallogr. **17** (1973), 1089.

13. W. J. Borer, Hs. H. Günthard, and P. Ballmer, *Solid State Reactions and Defects in Verneuil Laser Rubies,* Helv. Phys. Acta **43** (1970), 74.

14. G. Burns, E. A. Geiss, B. A. Jenkins, and M. I. Nathan, *Cr^{3+} Fluorescence in Garnets and Other Crystals,* Phys. Rev. **139** (1965), A1687.

15. L. D. Calvert, E. J. Gabe, and Y. Le Page, *Ruby Spheres for Aligning Single-Crystal Diffractometers,* Acta Crystallogr. **A37** (1981), C314.

16. R. J. Collins, D. F. Nelson, A. L. Schawlow, W. Bond, C.G.B. Garret, and W. Kaiser, *Coherence, Narrowing, Directionality, and Relaxation Oscillations in the Light Emission from Ruby,* Phys. Rev. Lett. **5** (1960), 303.

17. D. E. Cox, A. R. Moodenbaugh, A. W. Sleight, and H.-Y. Chen, *Structural Refinement of Neutron and X-ray Data by the Rietveld Method: Application to Al_2O_3 and $BiVO_4$,* National Bureau Standards (U.S.) Spec. Publ. No. 567 (1980), p 189.

18. J. M. Eggleston, L. G. DeShazer, and K. W. Kangas, *Characteristics and Kinetics of Laser-Pumped Ti:Sapphire Oscillators,* IEEE J. Quantum Electron. **QE-24** (1988), p 1009.

19. K. Eigenmann, K. Kurtz, and Hs. H. Günthard, *Solid State Reactions and Defects in Doped Verneuil Sapphire III. Systems α-AL_2O_3:Fe, α-Al_2O_3:Ti and α-Al_2O_3:(Fe, Ti),* Helv. Phys. Acta **45** (1972), 452.

20. W. M. Fairbank, G. K. Klauminzer, and A. L. Schawlow, *Excited-State Absorption in Ruby, Emerald, and MgO:Cr^{3+},* Phys. Rev. **B11** (1975), 60.

21. J. Ferguson and D. L. Wood, *Crystal Field Spectra of $d^{3,7}$ Ions: VI.—The Weak Field Formalism and Covalency,* Aust. J. Chem. **23** (1970), 861.

22. B. F. Gächter and J. A. Koningstein, *Zero Phonon Transitions and Interacting Jahn-Teller Phonon Energies from the Fluorescence Spectrum of α-Al_2O_3:Ti^{3+},* J. Chem. Phys. **60** (1974), 2003.

23. S. Geschwind, P. Kisleuk, M. P. Klein, J. P. Remeika, and D. L. Wood, *Sharp-Line Fluorescence, Electron Paramagnetic Resource, and Thermoluminescence of Mn^{4+} in α-Al_2O_3,* Phys. Rev. **126** (1962), 1684.

24. S. Geschwind and J. P. Remeika, *Spin Resonance of Transition Metal Ions in Corundum,* J. Appl. Phys. Suppl. **33** (1962), 370.

25. U. Hochli and K. A. Muller, *Observations of the Jahn-Teller Splitting of Three-Valent d^7 Ions Via Orbach Relaxation,* Phys. Rev. Lett. **12** (1964), 730.

26. N. S. Hush and J. M. Hobbs, *Absorption Spectra of Crystals Containing Transition Metal Ions,* Prog. Inorgan. Chem. **10** (1968), 259–486.

27. Yu. Zh. Isaenko, V. P. Puzikova, L. N. Raiskaya, and E. M. Spitsyn, *Laser with a Titanium-Activated Corundum Active Element and Acoustooptic Tuning of the Radiation Wavelength,* Sov. J. Quantum Electron. **18** (1989), 1258.

28. R. R. Joyce and P. L. Richards, *Far-Infrared Spectra of Al$_2$O$_3$ Doped with Ti, V, and Cr,* Phys. Rev. **179** (1969), 375.

29. W. Kaiser, S. Sugano, and D. L. Wood, *Splitting of the Emission Lines of Ruby by an External Electric Field,* Phys. Rev. Lett. **6** (1961), 605.

30. T. Kushida, *Absorption Spectrum of Optically Pumped Ruby: I.— Experimental Studies of Spectrum of Excited States,* J. Phys. Soc. Jpn. **21** (1966), 1331.

31. P. Lacovara and L. Esterowitz, *Growth, Spectroscopy and Lasing of Titanium-Doped Sapphire,* IEEE J. Quantum Electron. **QE-21** (1985), 1614.

32. R. Lacroix, U. Hochli, and K. A. Muller, *Strong Field G-Value Calculation for d^7 Ions in Octahedral Surroundings,* Helv. Phys. Acta **37** (1964), 627.

33. W. Low and E. L. Offenbacher, *Electron Spin Resonance of Magnetic Ions in Complex Oxides, Review of ESR Results in Rutile, Perovskite, Spinel, and Garnet Structures,* Solid State Physics **17** (1965), 135.

34. A. D. Liehr, *The Three Electron (or Hole) Cubic Ligand Field Spectrum,* J. Phys. Chem. **67** (1963), 1314.

35. A. Linz, Jr., and R. E. Newnham, *Ultraviolet Absorption Spectra in Ruby,* Phys. Rev. **123** (1961), 500.

36. R. Louat and E. Duval, *Temperature Dependence of the $^1A_1 \rightarrow ^1E(^1T_1)$ Zero-Phonon Transition of Co^{3+} in α-Al$_2$O$_3$,* Phys. Status Solidi **42** (1970), K93.

37. A. Lupei, V. Lupei, C. Ionescu, H. G. Tang, and M. L. Chen, *Spectroscopy of Ti^{3+}:α-Al$_2$O$_3$,* Optics Commun. **59** (1986), 36.

38. R. M. Macfarlane, *Perturbation Methods in the Calculation of Zeeman Interactions and Magnetic Dipole Line Strengths for d^3 Trigonal-Crystal Spectra,* Phys. Rev. **B1** (1970), 989.

39. R. M. Macfarlane, *On the Ground-State Splitting in Ruby,* J. Chem. Phys. **42** (1965), 442.

40. R. M. Macfarlane, *Optical and Magnetic Properties of Trivalent Vanadium Complexes,* J. Chem. Phys. **40** (1964), 373.

41. R. M. Macfarlane, *Analysis of the Spectrum of d^3 Ions in Trigonal Crystal Fields,* J. Chem. Phys. **39** (1963), 3118.

42. R. M. Macfarlane, J. Y. Wong, and M. D. Sturge, *Dynamic Jahn-Teller Effect in Octahedrally Coordinated d^1 Impurity Systems,* Phys. Rev. **166** (1968), 250.

43. T. H. Maiman, *Stimulated Optical Radiation in Ruby*, Nature **187** (1960), 493.

44. T. H. Maiman, *Optical Maser Action in Ruby*, Br. Commun. Electron. **7** (1960), 674.

45. R. Marshall, S. S. Mitra, P. J. Gielisse, J. N. Plendl, and L. C. Mansur, *Infrared Lattice Spectra of α-Al_2O_3 and Cr_2O_3*, J. Chem. Phys. **43** (1965), 2893.

46. Z. G. Mazurak, M. B. Czaja, J. Hanuza, and B. Jezowska-Trzebiatowska, *The Spectroscopy of Cr^{3+} Doped Natural Garnets and Emerald as Well as Synthetic Alexandrite and Corundum,* in Proc. First Int. School on Excited States of Transition Elements, B. Jezowska-Trzebiatowska, J. Legendziewicz, and W. Strek, eds., World Scientific (1989), p 331.

47. F. J. McClung, S. E. Schwarz, and F. J. Meyers, *R_2 Line Optical Maser Action in Ruby*, J. Appl. Phys. **33** (1962), 3139.

48. D. S. McClure, *Optical Spectra of Transition-Metal Ions in Corundum*, J. Chem. Phys. **36** (1962), 2757.

49. D. S. McClure, *Progress in Solid State Physics*, Academic Press, New York (1959), Vol. 9.

50. R. Moncorge, G. Boulon, D. Vivien, A. M. Lejus, R. Collongues, V. Djevahirdjian, K. Djevahirdjian, and R. Cagnard, *Optical Properties and Tunable Laser Action of Verneuil-Grown Single Crystals of Al_2O_3:Ti^{3+}*, IEEE J. Quantum Electron. **QE-24** (1988), 1049.

51. K. Moorjani and N. McAvoy, *Optical Spectra of Trivalent Iron in Trigonal Fields*, Phys. Rev. **132** (1963), 504.

52. P. F. Moulton, *Tunable Solid-State Lasers Targeted for a Variety of Applications*, Laser Focus **23** (1987), 56.

53. P. F. Moulton, *Spectroscopic and Laser Characteristics of Ti^{3+}:Al_2O_3*, J. Opt. Soc. Am. **B3** (1986), 125.

54. E. D. Nelson, J. Y. Wong, and A. L. Schawlow, *Far Infrared Spectra of Al_2O_3:Cr^{3+} and Al_2O_3:Ti^{3+}*, Phys. Rev. **156** (1967), 298.

55. E. D. Nelson, J. Y. Wong, and A. L. Schawlow, *Far Infrared Spectra of Al_2O_3:Cr^{3+} and Al_2O_3:Ti^{3+}* in *Optical Properties of Ions in Crystals*, H. M. Crosswhite and H. W. Moos, eds., Interscience, New York (1967), p 375.

56. R. E. Newnham and Y. M. de Haan, *Refinement of the α Al_2O_3, Ti_2O_3, V_2O_3, and Cr_2O_3 Structures*, Z. Kristallogr. Krystallgeom. Krystallphys. Kristallchem. **117** (1962), 235.

57. A. Pinto, *Doubled Titanium Sapphire as Tunable Visible Laser*, Laser Focus **23** (1987), 58.

58. M.H.L. Pryce and W. A. Runciman, *The Absorption Spectrum of Vanadium Corundum*, Discuss. Faraday Soc. **26** (1958), 34.

59. C. Reber and H. U. Güdel, *Near-Infrared Luminescence Spectroscopy of Al$_2$O$_3$:V^{3+} and YP$_3$O$_9$:V^{3+}*, Chem. Phys. Lett. **154** (1989), 425.

60. S. Sakatsume and I. Tsujikawa, *Sharp Absorption Lines of V^{3+}-Al$_2$O$_3$ in the Near Infrared Region*, J. Phys. Soc. Jpn. **19** (1964), 1080.

61. A. Sanchez, A. J. Strauss, R. L. Aggarwal, and R. E. Fahey, *Crystal Growth, Spectroscopy, and Laser Characteristics of Ti:Al$_2$O$_3$*, IEEE J. Quantum Electron. **QE-24** (1988), 995.

62. A. L. Schawlow and G. E. Devlin, *Simultaneous Optical Maser Action in Two Satellite Lines*, Phys. Rev. Lett. **6** (1961), 96.

63. P. A. Schulz, *Single-Frequency Ti:Al$_2$O$_3$ Ring Laser*, IEEE J. Quantum Electron. **QE-24** (1988), 1039.

64. B. K. Sevastyanov, Kh. S. Bagdasavov, E. A. Fedorov, V. B. Semenov, I. N. Tsigler, K. P. Chirkina, L. S. Starostina, A. P. Chirkin, A. A. Minaev, V. P. Orekhova, V. F. Seregin, and A. N. Kolerov, *Spectral and Lasing Characteristics of Corundum Crystals Activated by Ti^{3+} (Al$_2$O$_3$:Ti^{3+}) Ions*, Sov. Phys. Dokl. **30** (1986), 508.

65. B. K. Sevastyanov, Kh. S. Bagdasavov, E. A. Fedorov, V. B. Semenov, I. N. Tsigler, K. P. Chirkina, L. S. Starostina, A. P. Chirkin, A. A. Minaev, V. P. Orekhova, V. F. Seregin, A. N. Kolerov, and A. N. Vratskii, *Tunable Laser Based on Al$_2$O$_3$:Ti^{3+} Crystal*, Sov. Phys. Crystallogr. **29** (1984), 566.

66. R. R. Sharma and T. P. Das, *Crystalline Fields in Corundum-Type Lattices*, J. Chem. Phys. **41** (1964), 3581.

67. M. D. Sturge, F. R. Merritt, L. F. Johnson, H. J. Guggenheim, and J. P. van der Ziel, *Optical and Microwave Studies of Divalent Vanadium in Octahedral Fluoride Coordination*, J. Chem. Phys. **54** (1971), 405.

68. S. Sugano and M. Peter, *Effect of Configuration Mixing and Covalency on the Energy Spectrum of Ruby*, Phys. Rev. **122** (1961), 381.

69. S. Sugano and M. Shinada, *Theoretical Analysis of Absorption Spectrum of Optically Pumped Ruby*, in *Optical Properties of Ions in Crystals*, H. M. Crosswhite and H. W. Moos, eds., Interscience, New York (1967), p 187.

70. S. Sugano and Y. Tanabe, *The Line Spectra of Cr^{3+} Ion in Crystals*, Discuss. Faraday Soc. **26** (1958), 43.

71. D. T. Sviridov, R. K. Sviridova, N. I. Kulik, and V. B. Glasko, *Optical Spectra of the Isoelectronic Ions V^{2+}, Cr^{2+}, and Mn^{4+} in an Octahedral Coordination*, J. Appl. Spectrosc. **30** (1979), 334.

72. T. Tatsukawa, M. Inoue, and H. Yagi, *Fine Structure Constant at V-Band Region of Cr^{3+} in Ruby Crystal*, J. Phys. Soc. Jpn. **36** (1974), 908.

73. P. Thompson, D. E. Cox, and J. B. Hastings, *Rietveld Refinement of Debye-Scherrer Synchrotron X-Ray Data from Al_2O_3*, J. Appl. Crystallogr. **20** (1987), 79.

74. H. H. Tippins, *Charge-Transfer Spectra of Transition-Metal Ions in Corundum*, Phys. Rev. **B1** (1970), 126.

75. A. van Die, A. Leenaers, W. van der Weg, and G. Blasse, *A Search for Luminescence of the Trivalent Manganese Ion in Solid Aluminate*, Mater. Res. Bull. **22** (1987), 781.

76. K. F. Wall, R. L. Aggarwal, R. E. Fahey, and A. J. Strauss, *Small-Signal Gain Measurements in a $Ti:Al_2O_3$ Amplifier*, IEEE J. Quantum Electron. **QE-24** (1988), 1016.

77. A. Wasiela, D. Block, and Y. M. D'Aubigne, *Chromium-Gallium Complexes in Al_2O_3: II.—Energy Transfer*, J. Lumin. **36** (1986), 23.

78. A. Wasiela, Y. M. D'Aubigne, and D. Block, *Chromium-Gallium Complexes in Al_2O_3: I.—Luminescence*, J. Lumin. **36** (1986), 11.

79. J. Y. Wong, M. J. Berggren, and A. L. Schawlow, *Far-Infrared Spectrum of $Al_2O_3:V^{4+}$*, J. Chem. Phys. **49** (1968), 835.

80. E. J. Woodbury and W. K. Ng, *Ruby Laser Operation in the Near I.R.*, Proc. IRE **50** (1962), 2367.

81. Y. Y. Yeung and D. J. Newman, *Superposition-Model Analysis for the $Cr^{3+}\ ^4A_2$ Ground State*, Phys. Rev. **B34** (1986), 2258.

25. MgAl₂O₄

25.1 Crystallographic Data on MgAl₂O₄

Cubic O_h^7 $(Fd3m)$, 227, $Z = 8$

Ion	Site	Symmetry	x^a	y	z	q	$\alpha\,(\text{Å}^3)^b$
Mg	8(a)	T_d	0	0	0	2	0.0809
Al	16(d)	D_{3d}	5/8	5/8	5/8	3	0.053
O	32(e)	C_{3v}	x	x	x	–2	1.349

^{a}X-ray data: $a = 8.080$ Å, $x = 0.389$ (Wyckoff, 1968).
^{b}Schmidt et al (1979).

25.2 Crystal-Field Components, A_{nm} (cm⁻¹/Åⁿ)

25.2.1 For Al (D_{3d}) site (rotated so that z-axis is parallel to (111) crystallographic axis)

A_{nm}	Point charge	Self-induced	Dipole	Total
A_{20}	–2,283	–2008	23,088	18,797
A_{40}	–20,238	8618	–4,838	–16,458
A_{43}	–23,688	8781	–2,390	–17,297

25.2.2 For Mg (T_d) site

$A_{nm}{}^a$	Point charge	Self-induced	Dipole	Total
A_{32}	30,202	–8210	–9151	12,840
A_{40}	–13,597	5447	5350	–2,800

$^aA_{44} = \sqrt{5/14} = A_{40}$

25.3 Experimental Parameters (cm⁻¹) for Transition-Metal Ions

Ion	nd^N	Site	$F^{(2)}$	$F^{(4)}$	α	ζ	B_{20}	B_{40}	$B_{4m}{}^a$	Ref
Cr³⁺	$3d^3$	Al	56,700	40,320	—	250	4608	–30,625	–28,415	18
Cr³⁺	$3d^3$	Al	56,700	40,320	—	—	0	–25,550	–30,538	18
Fe²⁺	$3d^6$	Mg	—	—	—	392	0	–9,387	—	11
Co²⁺	$3d^7$	Mg	—	—	—	—	0	–8,400	—	3

^{a}For Al site $B_{4m} = B_{43}$, and for Mg site $B_{4m} = B_{44}$ $(B_{44} = \sqrt{5/14}\ B_{40})$

25.4   Bibliography and References

1. S. A. Basun, P. Deren, A. A. Kaplyanskii, W. Strek, and S. P. Feofilov, *Narrowing of Fluorescence Lines and Optical Detection of Nonequilibrium Terahertz Acoustic Phonons in Disordered MgAl$_2$O$_4$:Cr^{3+} Crystals*, Sov. Phys. Solid State **31** (1989), 460.

2. R. Clausen and K. Petermann, *Mn^{2+} and Fe^{3+} Doped Oxides for Short Wavelength Solid State Lasers*, J. Lumin. **40,41** (1988), 185.

3. R. D. Gillen and R. E. Solomon, *Optical Spectra of Chromium (III), Cobalt (II), and Nickel (II) Ions in Mixed Spinels*, J. Phys. Chem. **74** (1970), 4252.

4. M. R. Kokta, *Crystal Growth and Characterization of Oxide Host Crystals for Tunable Lasers*, in Proc. SPIE, Lasers and Nonlinear Optical Materials, L. G. DeShazer, ed., **681** (1986), 50.

5. W. Mikenda, *N-Lines in the Luminescence Spectra of Cr^{3+}-Doped Spinels (III) Partial Spectra*, J. Lumin. **26** (1981), 85.

6. W. Mikenda and A. Preisinger, *N-Lines in the Luminescence Spectra of Cr^{3+}-Doped Spinels: (I) Identification of N-Lines*, J. Lumin. **26** (1981), 53.

7. W. Mikenda and A. Preisinger, *N-Lines in the Luminescence Spectra of Cr^{3+}-Doped Spinels: (II) Origins of N-Lines*, J. Lumin. **26** (1981), 67.

8. R. Mlcak and A. H. Kitai, *Cathodoluminescence of Mn^{2+} Centers in MgAl$_2$O$_4$ Spinels*, J. Lumin. **46** (1990), 391.

9. R. Pappalardo, D. L. Wood, and R. C. Linares, Jr., *Optical Absorption Spectra of Ni-Doped Oxide Systems: I*, J. Chem. Phys. **35** (1961), 1460.

10. P. C. Schmidt, A. Weiss, and T. P. Das, *Effect of Crystal Fields and Self-Consistency on Dipole and Quadrupole Polarizabilities of Closed-Shell Ions*, Phys. Rev. **B19** (1979), 5525.

11. G. A. Slack, F. S. Ham, and R. M. Chrenko, *Optical Absorption of Tetrahedral Fe^{2+} (3d^6) in Cubic ZnS, CdTe, and MgAl$_2$O$_4$*, Phys. Rev. **152** (1966), 376.

12. W. Strek, P. Deren, and B. Jezowska-Trzebiatowska, *The Co^{2+} Doped MgAl$_2$O$_4$ Spinel: Potential Candidate for Tunable Solid State Lasers*, in Proc. First International School on Excited States of Transition Elements, B. Jezowska-Trzebiatowska, J. Legendziewicz, and W. Strek, eds., World Scientific, New Jersey (1989), p 490.

13. W. Strek, P. Deren, and B. Jezowska-Trzebiatowska, *Optical Properties of Cr^{3+} in MgAl$_2$O$_4$ Spinel*, Physica **B152** (1988), 380.

14. W. Strek, P. Deren, and B. Jezowska-Trzebiatowska, *The Nature of Cr(III) Luminescence in MgAl$_2$O$_4$ Spinel*, J. Lumin. **40,41** (1988), 421.

15. W. Strek, P. Deren, and B. Jezowska-Trzebiatowska, *Broad-Band Emission of Cr^{3+} in MgAl$_2$O$_4$ Spinel*, J. Phys. Paris **48** (1987), C7-475.

16. W. Strek, P. Deren, and B. Jezowska-Trzebiatowska, *Optical Properties of Ti^{3+} in MgAl$_2$O$_4$ Spinel*, J. Phys. Paris **48** (1987), C7-455.

17. H. A. Weakliem, *Optical Spectra of Ni^{2+}, Co^{2+}, and Cu^{2+}*, J. Chem. Phys. **36** (1962), 2117.

18. D. L. Wood, G. F. Imbusch, R. M. MacFarlane, P. Kisliuk, and D. M. Larkin, *Optical Spectrum of Cr^{3+} Ions in Spinels*, J. Chem. Phys. **48** (1968), 5255.

19. R.W.G. Wyckoff, *Crystal Structures*, vol 3, Interscience, New York (1968), p 75.

26. $A_3B_2Ge_3O_{12}$ (Germanium Garnet)

26.1 Crystallographic Data on $A_3B_2Ge_3O_{12}$

Cubic O_h^{10} (Ia3d), 230, Z = 8

Ion	Site	Symmetry	x	y	z	q	α (Å³)[a]
A	24(c)	D_2	0	1/4	1/8	2	α_A
B	16(a)	C_{3i}	0	0	0	3	α_B
Ge	24(d)	S_4	0	1/4	3/8	4	0.12
O	96(h)	C_1	x	y	z	-2	1.349

[a]Values for α are from Schmidt et al (1979), except for values not given there, for which the α values are from Fraga et al (1976).

26.2 X-Ray Data on $A_3B_2Ge_3O_{12}$

A	B	a (Å)	x	y	z	α_A (Å³)[a]	α_B (Å³)[a]	Ref
Ca	Al	12.118	−0.03345	0.0488	0.14753	0.564	0.0530	26
Ca	Ga	12.251	—	—	—	0.564	0.19	29
Ca	Cr	12.262	—	—	—	0.564	0.29	29
Ca	V	12.324	—	—	—	0.564	0.31	29
Ca	Fe	12.325	—	—	—	0.564	0.24	29
Ca	Sc	12.519	−0.0352	0.0524	0.1552	0.564	0.77	26
Ca	Lu	12.590	−0.03538	0.05607	0.15989	0.564	0.540	26
Ca	In	12.735	−0.0363	0.0543	0.15724	0.564	0.54	26
Sr	Sc	12.785	−0.03861	0.04909	0.15339	1.039	0.540	26
Cd	Sc	12.458	−0.03437	0.05300	0.15564	0.840	0.540	15
Cd	Al	12.08	—	—	—	0.840	0.0530	29
Cd	Cr	12.20	—	—	—	0.840	0.29	29
Cd	Fe	12.26	—	—	—	0.840	0.24	29
Cd	Ga	12.19	—	—	—	0.840	0.19	29
Mn	Al	11.902	—	—	—	0.460	0.0530	29
Mn	Cr	12.027	—	—	—	0.460	0.29	29
Mn	Fe	12.087	—	—	—	0.460	0.24	29
Mn	Ga	12.00	—	—	—	0.460	0.19	29

[a]Values for α are from Schmidt et al (1979), except for values not given there, for which the α values are from Fraga et al (1976).

26.3 Crystal-Field Components, A_{nm} (cm^{-1}/Å^n)

Rotated so that z-axis is parallel to (111) crystallographic axis

26.3.1 For Al ion in 16(a) (C_{3i}) site in $Ca_3Al_2Ge_3O_{12}$

A_{nm}	Point charge	Self-induced	Dipole	Total
A_{20}	11,161	−451.6	−35,565	−24,855
A_{40}	−20,669	8255	9,602	−2,812
ReA_{43}	2,625	−1473	7,554	8,706
ImA_{43}	−22,662	9272	5,206	−8,184
$|A_{43}|$	22,814	—	—	11,949

26.3.2 For Sc ion in 16(a) (C_{3i}) site in $Ca_3Sc_2Ge_3O_{12}$

A_{nm}	Point charge	Self-induced	Dipole	Total
A_{20}	9,208	−299.1	−28,304	−19,395
A_{40}	−13,449	4192	5,733	−3,523
ReA_{43}	1,227	−651.0	4,588	5,164
ImA_{43}	−14,541	4678	2,550	−7,313
$|A_{43}|$	14,593	—	—	8,953

26.3.3 For Lu ion in 16(a) (C_{3i}) site in $Ca_3Lu_2Ge_3O_{12}$

A_{nm}	Point charge	Self-induced	Dipole	Total
A_{20}	10,201	−354.1	−26,892	−17,044
A_{40}	−11,205	3133	4,153	−3,920
ReA_{43}	525.3	−376.6	3,896	4,044
ImA_{43}	−11,743	3417	1,241	−7,085
$|A_{43}|$	11,754	—	—	8,158

26.3.4 For In ion in 16(a) (C_{3i}) site in $Ca_3In_2Ge_3O_{12}$

A_{nm}	Point charge	Self-induced	Dipole	Total
A_{20}	8,212	−237.9	−25,117	−17,143
A_{40}	−11,352	3214	4,616	−3,523
ReA_{43}	612.7	−398.0	3,909	4,123
ImA_{43}	−12,272	3594	1,839	−6,838
$|A_{43}|$	12,287	—	—	7,985

26.3.5 For Sc ion in 16(a) (C_{3i}) site in $Sr_3Sc_2Ge_3O_{12}$

A_{nm}	Point charge	Self-induced	Dipole	Total
A_{20}	2,456	79.17	−22,887	−20,343
A_{40}	−12,168	3703	6,298	−2,167
ReA_{43}	1,424	−667.5	4,090	4,847
ImA_{43}	−14,278	4414	3,785	−6,079
$|A_{43}|$	14,349	—	—	7,775

26.3.6 For Sc ion in 16(a) (C_{3i}) site in $Cd_3Sc_2Ge_3O_{12}$

A_{nm}	Point charge	Self-induced	Dipole	Total
A_{20}	10,782	−444.2	−29,570	−19,232
A_{40}	−13,735	4323	5,561	−3,851
ReA_{43}	1,236	−664.4	4,670	5,241
ImA_{43}	−14,599	4729	2,266	−7,605
$|A_{43}|$	14,652	—	—	9,236

26.4 Experimental Values (cm^{-1}) of B_{40}, $F^{(2)}$, and $F^{(4)}$ for nd^N Ions

Ion	nd^N	$F^{(2)}$	$F^{(4)}$	$B_{40}{}^a$	Ref.
Cr^{3+}	$3d^3$	66,121	31,102	−21,200	16[b]
Mn^{2+}	$3d^5$	52,570	47,880	−13,160	27[c]

$^aB_{43} = \sqrt{10/7}\, B_{40}$
^{b}In $Ca_3Ga_2Ge_3O_{12}$
^{c}In $Cd_{3-x}Ca_xAl_2Ge_3O_{12}$

26.5 Bibliography and References

1. A. M. Antyukhov, A. A. Sidorov, I. A. Ivanov, and A. V. Antovov, *The Thermal Expansion Coefficients of Crystals of Certain Garnets over the Range 6–310K*, Neorg. Mat. **23** (1987), 702.

2. K. P. Belov, D. G. Mamsurova, B. V. Mill, and V. I. Sokolov, *Ferromagnetism of the Garnet $Mn_3Cr_2Ge_3O_{12}$*, Zh. Eksp. Teor. Fiz. Pis'ma Red. **16** (1972), 173.

3. K. P. Belov, B. V. Mill, V. I. Sokolov, and O. I. Shevaleevskii, *Antiferromagnetic Resonance in the Garnet $Ca_3Fe_2Ge_3O_{12}$*, JETP Lett. **20** (1974), 42.

4. Th. Brückel, Y. Chernenkov, B. Dorner, V. P. Plakhty, and O. P. Smirnov, *Spin Wave Spectrum and Exchange Interactions in the Antiferromagnetic Garnet $Ca_3Cr_2Ge_3O_{12}$ by Inelastic Neutron Scattering*, Z. Phys. **B79** (1990), 389.

5. S. Fraga, J. Karwowski, and K.M.S. Saxena, *Handbook of Atomic Data*, vol 5 (1976), p 319.

6. S. Geller and C. E. Miller, appendix by R. G. Treuting, *New Synthetic Garnets*, Acta Crystallogr. **13** (1960), 179.

7. R. Gieniusz and A. Maziewski, *Low Power Light Deflector on Domain Structure in $(Y,Ca)_3(Fe,Ge)_5O_{12}$*, Opt. Commun. **51** (1984), 167.

8. V. Havlicek, P. Novak, and B. V. Mill, *EPR of V^{4+} Ion in Several Garnet Systems*, Phys. Status Solidi (b)**64** (1974), K19.

9. A. A. Kaminskii, B. V. Mill, and A. V. Butashin, *New Possibilities for Exciting Stimulated Emission in Inorganic Crystalline Materials with the Garnet Structure*, Inorg. Mater. USSR **19** (1983), 2056.

10. Z. A. Kazei, B. V. Mill, and V. I. Sokolov, *Peculiarities of the Metamagnetic Transition of a Garnet Single Crystal $Ca_3Mn_2Ge_3O_{12}$*, JETP Lett. **31** (1980), 308.

11. Z. A. Kazei, B. V. Mill, and V. I. Sokolov, *Cooperative Jahn-Teller Effect in the Garnet $Ca_3Mn_2Ge_3O_{12}$*, JETP Lett. **24** (1976), 203.

12. V. F. Kitaeva, E. V. Zharikov, and I. L. Chistyi, *The Properties of Crystals with Garnet Structure*, Phys. Status Solidi (a)**92** (1985), 475.

13. G. S. Krinchik, V. D. Gorbunova, V. S. Gushchin, and B. V. Mill, *Absorption of Light in One-Sublattice Iron Garnets*, Sov. Phys. Solid State **22** (1980), 156.

14. M. D. Lind and S. Geller, *Crystal Structure of the Garnet $\{Mn_3\}[Fe_2](Ge_3)O_{12}$*, Z. Kristallogr. **129** (1969), 427.

15. B. V. Mill, E. L. Belokoneva, M. A. Simonov, and N. V. Belov, *Refined Crystal Structures of the Scandium Garnets $Ca_3Sc_2Si_3O_{12}$, $Ca_3Sc_2Ge_3O_{12}$, and $Cd_3Sc_2Ge_3O_{12}$*, Zh. Strukt. Khim. **18** (1977), 399.

16. A. E. Nosenko, A. I. Bilyi, L. V. Kostyk, and V. V. Kravchishin, *Emission Spectra of Cr^{3+} Ions in $Ca_3Ga_2Ge_3O_{12}$ Single Crystals*, Opt. Spectrosc. **57** (1984), 510.

17. A. E. Nosenko, B. V. Padlyak, and V. V. Kravchishin, *Electron Spin Resonance of Mn^{4+} Ions in $Ca_3Ga_2Ge_3O_{12}$ Single Crystals*, Sov. Phys. Solid State **27** (1985), 2083.

18. P. Novak, V. Havlicek, B. V. Mill, V. I. Sokolov, and O. I. Shevaleevskii, *EPR of Fe^{3+} Ions in Several Germanate Garnets and Magnetocrystalline Anisotropy of $Ca_3Fe_2Ge_3O_{12}$*, Solid State Commun. **19** (1976), 631.

19. A. E. Nosenko, A. P. Abramov, L. V. Kostyk, A. I. Bilyi, and V. V. Kravchishin, *Growth and Luminescence Properties of $Ca_3Ga_2Ge_3O_{12}$:Mn^{4+} Single Crystals*, Opt. Spectrosc. (USSR) **61** (1986), 648.

20. A. Pajaczkowska, G. Jasiolek, and K. Majcher, *Crystal Growth and Some Structural Investigations of $Mn_3A_2Ge_3O_{12}$ (MAGG) Garnets, Where A = Cr, Fe, Ga*, J. Cryst. Growth **79** (1986), 417.

21. R. Pappalardo, D. L. Wood, and R. C. Linares, Jr., *Optical Absorption Spectra of Ni-Doped Oxide Systems: I*, J. Chem. Phys. **35** (1961), 1460.

22. R. Plumier and M. Sougi, *Neutron Diffraction Study in Magnetic Field of Antiferromagnetic Garnet $Ca_3Mn_2Ge_3O_{12}$*, J. Phys. Paris Lett. **40** (1979), L-213.

23. P. C. Schmidt, A. Weiss, and T. P. Das, *Effect of Crystal Fields and Self-Consistency on Dipole and Quadrupole Polarizabilities of Closed-Shell Ions*, Phys. Rev. **B19** (1979), 5525.

24. V. I. Sokolov and O. I. Shevaleevskii, *Antiferromagnetic Resonance in the Cubic Crystal FeGeG and CrGeG*, Sov. Phys. JETP **45** (1977), 1245.

25. T. V. Valyanskaya, B. V. Mill, and V. I. Sokolov, *Antiferromagnetic Ordering of Cr^{3+} in Octahedral Garnet Sublattice*, Sov. Phys. Solid State **18** (1976), 696.

26. I. Veltrusky, *Effects of Covalency on Magnetic Properties of Tetrahedral V^{4+} Ion in Garnets*, Czech. J. Phys. **B28** (1978), 675.

27. J. Wiehl, R. Hirvle, W. Wischert, and S. Kemmler-Sack, *Cathodo- and Photoluminescence in the Mn^{2+} Activated Garnets $Cd_{3-x}Ca_xAl_2Ge_3O_{12}$*, Phys. Status Solidi **(a)111** (1989), 315.

28. D. Wolf and S. Kemmler-Sack, *Cathodo- and Photo-luminescence in Rare Earth Activated Garnets of Type $Ca_3M_2Ge_{3-x}Si_xO_{12}$ (M = Al, Ga, Sc, Y)*, Phys. Status Solidi **(a)98** (1986), 567.

29. R.W.G. Wyckoff, *Crystal Structures*, vol 3, Interscience, New York (1968), p 223.

27. ZnGa$_2$O$_4$

27.1 Crystallographic Data on ZnGa$_2$O$_4$

Cubic O_h^7 ($Fd3m$), 227, $Z = 8$

Ion	Site	Symmetry	x^a	y	z	q	α (Å^3)^b
Zn	8(a)	T_d	0	0	0	2	0.676
Ga	16(d)	D_{3d}	5/8	5/8	5/8	3	0.458
O	32(e)	C_{3v}	x	x	x	−2	1.349

^{a}X-ray data: $a = 8.330$ Å, $x = 0.38675$ (Hornstra and Keulen, 1972).
^{b}Schmidt et al (1979).

27.2 Crystal-Field Components, A_{nm} (cm^{-1}/Å^n)

27.2.1 For Zn (T_d) site

$A_{nm}{}^a$	Monopole	Self-induced	Dipole	Total
A_{32}	26,942	−6727	−7790	12,425
A_{40}	−11,793	4333	4419	−3,040

$^a A_{44} = \sqrt{5/14}\ A_{40}$

27.2.2 For Ga (D_{3d}) site (rotated so that z-axis is parallel to (111) crystallographic axis)

A_{nm}	Monopole	Self-induced	Dipole	Total
A_{20}	−2,616	−1736	19,928	15,576
A_{40}	−17,273	6713	−3,972	−14,531
A_{43}	20,288	−6815	1,969	15,442

27.3 Bibliography and References

1. B.H.T. Chai, *Final Report, New Laser Materials January 1985– December 1985*, Allied-Signal Corporation, Morristown, New Jersey (1986).

2. G.G.P. van Gorkom, J.C.M. Henning, and R. P. van Stapele, *Optical Spectra at Cr^{3+} Pairs in the Spinel ZnGa$_2$O$_4$*, Phys. Rev. **B8** (1973), 955.

3. J.C.M. Henning, *Weak Exchange Interactions in Chromium Doped ZnGa$_2$O$_4$*, Phys. Lett. **34A** (1971), 215.

4. J.C.M. Henning, J. H. den Boef, and G.G.P. van Gorkom, *Electron-Spin-Resonance Spectra of Nearest-Neighbor Cr^{3+} Pairs in the Spinel ZnGa$_2$O$_4$*, Phys. Rev. **B7** (1973), 1825.

5. J.C.M. Henning and J.P.M. Damen, *Exchange Interactions Within Nearest-Neighbor Cr^{3+} Pairs in Chromium-Doped Spinel ZnGa$_2$O$_4$*, Phys. Rev. **B3** (1971), 3852.

6. J. Hornstra and E. Keulen, *The Oxygen Parameter of the Spinel ZnGa$_2$O$_4$*, Philips Res. Rep. **27** (1972), 76.

7. O. Kahan and R. M. Macfarlane, *Optical and Microwave Spectra of Cr^{3+} in the Spinel ZnGa$_2$O$_4$*, J. Chem. Phys. **54** (1971), 5197.

8. W. Mikenda, *N-Lines in the Luminescence Spectra of Cr^{3+}-Doped Spinels: III.—Partial Spectra*, J. Lumin. **26** (1981), 85–98.

9. W. Mikenda and A. Preisinger, *N-Lines in the Luminescence Spectra of Cr^{3+}-Doped Spinels: I.—Identification of N-Lines*, J. Lumin. **26** (1981), 53-66.

10. W. Mikenda and A. Preisinger, *N-Lines in the Luminescence Spectra of Cr^{3+}-Doped Spinels: II.—Origins of N-Lines*, J. Lumin. **26** (1981), 67–83.

11. P. C. Schmidt, A. Weiss, and T. P. Das, *Effect of Crystal Fields and Self-Consistency on Dipole and Quadrupole Polarizabilities of Closed-Shell Ions*, Phys. Rev. **B19** (1979), 5525.

28. Cs₂GeF₆

28.1 Crystallographic Data on Cs₂GeF₆

Cubic O_h^5 (Fm3m), 225, $Z = 4$

Ion	Site	Symmetry	x^a	y	z	q	α (Å³)
Ge	4(a)	O_h	0	0	0	4	0.120[b]
Cs	8(c)	T_d	1/4	1/4	1/4	1	2.492[c]
F	24(e)	C_{4v}	x	0	0	−1	0.731[c]

[a]X-ray data: $a = 9.021$ Å, $x = 0.20$ (Wyckoff, 1968).
[b]Fraga et al (1976).
[c]Schmidt et al (1979).

28.2 Crystal-Field Components, A_{nm} (cm⁻¹/Åⁿ)

28.2.1 For Ge (O_h) site

$A_{nm}{}^a$	Monopole	Self-induced	Dipole	Total
A_{40}	21,689	−10,895	25,645	36,439

[a]$A_{44} = \sqrt{5/14}\ A_{40}$

28.2.2 For Cs (T_d) site

$A_{nm}{}^a$	Monopole	Self-induced	Dipole	Total
A_{32}	1162.7	139.27	−2378.4	−1076.4
A_{40}	−181.65	43.06	−174.48	−313.07

[a]$A_{44} = \sqrt{5/14}\ A_{40}$

28.3 Experimental Parameters (cm⁻¹)

Ion	nd^N	$F^{(2)}$	$F^{(4)}$	ζ	$B_{40}{}^a$	Ref
Mn⁴⁺	$3d^3$	52,794	50,929	380	45,885	2
Mn⁴⁺	$3d^3$	58,489	48,460	363	45,780	19
Re⁴⁺	$5d^3$	40,299	22,617	2953	73,143	10
Re⁴⁺	$5d^3$	40,019	22,554	3094	69,384	11
Os⁴⁺	$5d^4$	51,408	37,409	2800	51,450	17
Ir⁴⁺	$5d^5$	53,389	39,564	3500	51,450	17
Pt⁴⁺	$5d^6$	35,133	28,400	3579	66,150	15

$B_{44} = \sqrt{5/14}\ B_{40}$

28.4 Index of Refraction

$N = 1.3920 + 2.26\upsilon^2 \times 10^{-11}$ (υ in cm⁻¹) (Levchishina and Yamshchikov, 1982)

28.5   Bibliography and References

1. C. Campochiaro, D. S. McClure, P. Rabinowitz, and S. Dougal, *Two-Photon Spectroscopy of the $^4A_{2g} \rightarrow {}^4T_{2g}$ Transition in Mn^{4+} Impurity Ions in Crystals,* in *Vibronic Processes in Inorganic Chemistry,* C. D. Flint, ed., Kluwer Academic Publishers, Boston (1989), p 255.

2. R. -L. Chien, J. M. Berg, D. S. McClure, P. Rabinowitz, and B. N. Perry, *Two-Photon Electronic Spectroscopy of $Cs_2GeF_6:Mn^{4+}$,* J. Chem. Phys. **84** (1986), 4168.

3. S. L. Chodos, A. M. Black, and C. D. Flint, *Vibronic Spectra and Lattice Dynamics of Cs_2MnF_6 and $A_2^IM^{IV}F_6:MnF_6^{2-}$,* J. Chem. Phys. **65** (1976), 4816.

4. P. B. Dorain, *The Spectra of Re^{4+} in Cubic Crystal Fields,* in *Transition Metal Chemistry,* vol 4, R. L. Carlin, ed., Marcel Dekker, New York (1968), p 1.

5. S. Fraga, K.M.S. Saxena, and J. Karwowski, *Handbook of Atomic Data,* Elsevier, New York (1976).

6. L. Helmholz and M. E. Russo, *Spectra of Manganese (IV) Hexafluoride Ion (MnF_6^-) in Environments of O_h and D_{3d} Symmetry,* J. Chem. Phys. **59** (1973), 5455.

7. L. Kolditz, W. Wilde, and W. Hilmer, *Phase Determinations by X-Ray on the Thermal Dissociation and the Hydrolysis of Alkali Hexafluoro-germanates at Higher Temperatures,* Z. Anorg. Allg. Chem. **512** (1984), 48.

8. M. P. Laurent, H. H. Patterson, W. Pike, and H. Engstrom, *Pure and Mixed Crystal Optical Studies of the Jahn-Teller Effect for the d^6 Hexafluoroplatinate (IV) Ion,* Inorg. Chem. **20** (1981), 372.

9. T. F. Levchishina and E. F. Yamshchikov, *Optical Properties of Some Hexafluoro Complex Compounds,* Inorg. Mater. **18** (1982), 570.

10. J. LoMenzo, H. Patterson, S. Strobridge, and H. Engstrom, *Sharp-Line Absorption, Luminescence, Raman Studies for the $5d^3$ Hexafluororhenate (IV) Ion in Pure and Host Crystal Environments,* Mol. Phys. **40** (1980), 1401.

11. J. A. LoMenzo, S. Strobridge, and H. Patterson, *Electronic Absorption Spectra of the $5d^3$ Hexafluororhenate (IV) Ion,* J. Mol. Spectrosc. **66** (1977), 150.

12. P. A. Lund, *An Analysis of the Low Temperature Zeeman Spectra of $K_2GeF_6:Mn(IV)$ and $Cs_2GeF_6:Mn(IV)$,* Ph.D. thesis, Washington University (1977), Univ. Microfilm 77-32, 552.

13. N. B. Manson, Z. Hasan, and C. D. Flint, *Jahn-Teller Effect in the $^4T_{2g}$ State of Mn^{4+} in Cs_2SiF_6,* J. Phys. **C12** (1979), 5483.

14. J. Mrozack, *An Analysis of the EPR Spectra of Technetium (IV) Doped Potassium Germanium Hexafluoride and Cesium Germanium Hexafluoride,* Ph.D. thesis, Washington University (1978), Univ. Microfilm 7904198.

15. H. H. Patterson, W. J. DeBerry, J. E. Byrne, M. T. Hsu, and J. A. LoMenzo, *Low-Temperature Luminescence and Absorption Spectra of the d^6 Hexafluoroplatinate (IV) Ions Doped in a Cs$_2$GeF$_6$-Type Host Lattice,* Inorg. Chem. **16** (1977), 1698.

16. P. C. Schmidt, A. Weiss, and T. P. Das, *Effect of Crystal Fields and Self-Consistency on Dipole and Quadrupole Polarizabilities of Closed-Shell Ions,* Phys. Rev. **B19** (1979), 5525.

17. L. C. Weiss, P. J. McCarthy, J. P. Jasinski, and P. N. Schatz, *Absorption and Magnetic Circular Dichroism Spectra of Hexafluoroosmate (IV) and Hexafluoroiridate (IV) in the Cubic Host Cs$_2$GeF$_6$,* Inorg. Chem. **17** (1978), 2689.

18. R.W.G. Wyckoff, *Crystal Structures,* vol 3, Interscience, New York, (1968), p 339.

19. W. C. Yeakel, R. W. Schwartz, H. G. Brittain, J. L. Slater, and P. N. Schatz, *Magnetic Circularly Polarized Emission and Magnetic Circular Dichroism of Resolved Vibronic Lines in Cs$_2$GeF$_6$:Mn^{4+},* Mol. Phys. **32** (1976), 1751.

29. $R_2Ti_2O_7$ ($R \neq Y$)

29.1 Crystallographic Data on $R_2Ti_2O_7$

Cubic O_h^7 ($Fd3m$), 227 (second setting), Z = 8

Ion	Site	Symmetry	x	y	z	q	α $(\text{Å}^3)a$
R	16(c)	D_{3d}	0	0	0	3	a_R
Ti	16(d)	D_{3d}	1/2	1/2	1/2	4	0.220
O_1	8(a)	T_d	1/8	1/8	1/8	−2	1.349
O_2	48(f)	C_{2v}	x	1/8	1/8	−2	1.349
X	8(b)	T_d	3/8	3/8	3/8	—	—

^{a}Schmidt et al (1979).

29.2 X-Ray Data on $R_2Ti_2O_7$ (Knop et al, 1969) and Polariz-abilities, α_R, of Rare-Earth Ions, R^{3+} ($4f^N$) (Fraga et al, 1976)

N	R	a (Å)	x	α_R (Å^3)
5	Sm	10.2303	0.4230	1.11
6	Eu	10.1988	—	1.06
7	Gd	10.1857	0.4263	1.01
8	Tb	10.1560	—	0.97
9	Dy	10.1245	—	0.94
10	Ho	10.0979	—	0.90
11	Er	10.0759	0.4194	0.86
12	Tm	10.0533	—	0.83
13	Yb	10.0309	0.4201	0.80
14	Lu	10.0258	—	0.77

29.3 Crystal-Field Components, A_{nm}

Crystal-field components were obtained for R = Sm, Gd, Er, and Yb. A least-squares polynomial fit was used to obtain the components for the entire range of R. In 29.3.1 to 29.3.12, the crystal is rotated so that the z-axis is parallel to the (111) crystallographic axis.

29.3.1 A_{20} (cm^{-1}/Å^2) for R site 16c (D_{3d})

R	Monopole	Self-induced	Dipole	Total
Sm	8047	−1158	396.6	8759
Eu	8118	−1169	392.1	6299
Gd	8180	−1176	387.7	4447
Tb	8233	−1179	383.5	3203
Dy	8277	−1179	379.5	2569
Ho	8312	−1175	375.7	2543
Er	8339	−1167	372.1	3126
Tm	8356	−1155	368.7	4317
Yb	8364	−1139	365.4	6118
Lu	8364	−1120	362.3	8527

29.3.2 A_{40} (cm^{-1}/Å^4) for R site 16c (D_{3d})

R	Monopole	Self-induced	Dipole	Total
Sm	8,295	−2798	22.65	8132
Eu	9,564	−2854	27.27	8213
Gd	10,594	−2913	34.54	8296
Tb	11,384	−2975	44.46	8382
Dy	11,934	−3038	57.03	8470
Ho	12,245	−3105	72.26	8560
Er	12,316	−3175	90.14	8652
Tm	12,148	−3247	110.7	8747
Yb	11,740	−3321	133.9	8843
Lu	11,092	−3399	159.7	8942

29.3.3 A_{43} (cm^{-1}/Å^4) for R site 16c (D_{3d})

R	Monopole	Self-induced	Dipole	Total
Sm	2379	−570	−1038	771
Eu	2405	−580	−1055	769
Gd	2454	−596	−1088	770
Tb	2526	−617	−1135	775
Dy	2621	−642	−1196	782
Ho	2738	−673	−1273	792
Er	2877	−708	−1365	805
Tm	3040	−748	−1471	821
Yb	3225	−793	−1592	839
Lu	3433	−843	−1728	861

$R_2Ti_2O_7$

29.3.4 A_{60} (cm^{-1}/Å^6) for R site 16c (D_{3d})

R	Monopole	Self-induced	Dipole	Total
Sm	1750	−841	−56	853
Eu	1782	−863	−59	860
Gd	1815	−886	−63	866
Tb	1851	−909	−69	872
Dy	1887	−934	−76	878
Ho	1926	−959	−84	882
Er	1967	−985	−95	886
Tm	2009	−1012	−106	890
Yb	2052	−1040	−119	893
Lu	2098	−1069	−134	895

29.3.5 A_{63} (cm^{-1}/Å^6) for R site 16c (D_{3d})

R	Monopole	Self-induced	Dipole	Total
Sm	−698	198	182	−317
Eu	−710	204	186	−320
Gd	−726	211	193	−322
Tb	−747	221	202	−323
Dy	−771	232	214	−324
Ho	−800	245	229	−325
Er	−833	261	247	−326
Tm	−870	278	267	−326
Yb	−912	297	290	−325
Lu	−958	318	315	−325

29.3.6 A_{66} (cm^{-1}/Å^6) for R site 16c (D_{3d})

R	Monopole	Self-induced	Dipole	Total
Sm	1014	−307	−148	477
Eu	1031	−315	−151	641
Gd	1052	−324	−156	736
Tb	1077	−336	−163	762
Dy	1105	−350	−172	720
Ho	1137	−366	−183	608
Er	1172	−384	−196	428
Tm	1211	−403	−212	179
Yb	1254	−425	−229	−138
Lu	1300	−449	−248	−525

29.3.7 A_{20} (cm^{-1}/Å^2) for Ti site 16d (D_{3d})

R	Monopole	Self-induced	Dipole	Total
Sm	−13,382	1444	12,981	1043
Eu	−13,282	1442	13,035	1196
Gd	−13,476	1462	13,172	1158
Tb	−13,965	1504	13,391	930
Dy	−14,748	1568	13,692	513
Ho	−15,826	1655	14,076	−95
Er	−17,199	1764	14,542	−892
Tm	−18,866	1896	15,090	−1880
Yb	−20,828	2049	15,720	−3058
Lu	−23,084	2226	16,432	−4425

29.3.8 A_{40} (cm^{-1}/Å^4) for Ti site 16d (D_{3d})

R	Monopole	Self-induced	Dipole	Total
Sm	−15,449	5539	−4263	−14,173
Eu	−15,682	5671	−4303	−14,313
Gd	−15,823	5761	−4388	−14,450
Tb	−15,872	5809	−4520	−14,584
Dy	−15,830	5814	−4698	−14,714
Ho	−15,696	5778	−4922	−14,841
Er	−15,471	5699	−5193	−14,965
Tm	−15,153	5577	−5509	−15,086
Yb	−14,744	5414	−5872	−15,203
Lu	−14,244	5208	−6281	−15,317

29.3.9 A_{43} (cm^{-1}/Å^4) for Ti site 16d (D_{3d})

R	Monopole	Self-induced	Dipole	Total
Sm	21,584	−8272	2159	15,464
Eu	21,851	−8445	2186	15,596
Gd	22,079	−8593	2245	15,743
Tb	22,268	−8716	2337	15,906
Dy	22,417	−8814	2462	16,084
Ho	22,527	−8886	2619	16,279
Er	22,597	−8934	2808	16,489
Tm	22,628	−8956	3030	16,715
Yb	22,619	−8953	3285	16,956
Lu	22,571	−8925	3572	17,213

29.3.10 A_{60} (cm^{-1}/Å^6) for Ti site 16d (D_{3d})

R	Monopole	Self-induced	Dipole	Total
Sm	3912	−2792	30	1150
Eu	3968	−2854	33	1147
Gd	4026	−2914	47	1159
Tb	4088	−2973	73	1187
Dy	4151	−3032	110	1230
Ho	4217	−3088	159	1288
Er	4287	−3144	219	1361
Tm	4359	−3199	290	1450
Yb	4433	−3252	373	1554
Lu	4510	−3305	467	1673

29.3.11 A_{63} (cm^{-1}/Å^6) for Ti site 16d (D_{3d})

R	Monopole	Self-induced	Dipole	Total
Sm	−151	−93	1199	1039
Eu	−133	−26	1213	1054
Gd	−144	−20	1232	1068
Tb	−184	8	1255	1080
Dy	−252	59	1284	1090
Ho	−349	132	1317	1100
Er	−475	228	1355	1108
Tm	−629	347	1397	1115
Yb	−813	488	1445	1120
Lu	−1025	651	1497	1124

29.3.12 A_{66} (cm^{-1}/Å^6) for Ti site 16d (D_{3d})

R	Monopole	Self-induced	Dipole	Total
Sm	2346	−1666	−71	610
Eu	2385	−1706	−71	608
Gd	2426	−1747	−70	609
Tb	2470	−1789	−68	614
Dy	2517	−1830	−65	621
Ho	2566	−1874	−60	633
Er	2618	−1915	−56	648
Tm	2674	−1957	−50	666
Yb	2731	−2000	−43	688
Lu	2792	−2043	−36	713

29.3.13 A_{32} (cm^{-1}/Å^3) for vacancy site, X, $8b$ (T_d)

R	Monopole	Self-induced	Dipole	Total
Sm	−36,871	−539	606	−36,804
Eu	−37,235	−550	612	−37,174
Gd	−37,617	−561	626	−37,553
Tb	−38,017	−571	646	−37,944
Dy	−38,435	−582	674	−38,344
Ho	−38,871	−592	709	−38,756
Er	−39,325	−602	751	−39,177
Tm	−39,798	−611	801	−39,610
Yb	−40,288	−621	858	−40,053
Lu	−40,796	−630	922	−40,506

29.3.14 $A_{40}{}^a$ (cm^{-1}/Å^4) for vacancy site, X, $8b$ (T_d)

R	Monopole	Self-induced	Dipole	Total
Sm	34,671	−6814	4,550	33,401
Eu	35,195	−7003	3,408	33,797
Gd	35,598	−7125	2,704	34,162
Tb	35,877	−7180	2,439	34,496
Dy	36,035	−7168	26,13	34,798
Ho	36,071	−7089	3,225	35,069
Er	35,984	−6943	4,275	35,308
Tm	35,775	−6730	5,764	35,517
Yb	35,444	−6450	7,691	35,694
Lu	34,991	−6103	10,057	35,839

$^aA_{44} = \sqrt{5/14}\ A_{40}$

29.3.15 $A_{60}{}^a$ (cm^{-1}/Å^6) for vacancy site, X, $8b$ (T_d)

R	Monopole	Self-induced	Dipole	Total
Sm	−495	−807	406	−897
Eu	−496	−835	412	−919
Gd	−508	−854	420	−942
Tb	−530	−864	429	−965
Dy	−561	−865	438	−987
Ho	−602	−856	499	−1010
Er	−654	−840	461	−1033
Tm	−715	−814	474	−1056
Yb	−786	−780	487	−1078
Lu	−867	−737	502	−1101

$^aA_{64} = -\sqrt{7/2}\ A_{60}$

29.3.16 $A_{72}{}^a$ (cm^{-1}/Å^7) for vacancy site, X, $8b$ (T_d)

R	Monopole	Self-induced	Dipole	Total
Sm	1034	57	–0.3	1091
Eu	1055	59	–0.3	1114
Gd	1076	60	–0.3	1137
Tb	1098	62	–0.2	1160
Dy	1121	64	–0.2	1184
Ho	1143	66	–0.2	1209
Er	1167	67	–0.2	1234
Tm	1190	69	–0.2	1259
Yb	1215	71	–0.1	1286
Lu	1239	73	–0.1	1312

$^aA_{76} = \sqrt{11/13}\ A_{72}$

29.4 Experimental Parameters (cm^{-1}) (Antonov and Arsenev, 1976)

nd^N	Ion	$F^{(2)}$	$F^{(4)}$	$B_{40}{}^a$
$3d^2$	V^{3+}	63,903	41,378	–24,500
$3d^3$	Cr^{3+}	65,450	42,840	–26,222
$3d^3$	Mn^{4+}	—	—	–31,920
$3d^4$	Mn^{3+}	—	—	–28,406
$3d^6$	Co^{3+}	58,100	34,902	–24,360
$3d^7$	Co^{2+}	66,220	43,344	–8,442
$3d^7$	Ni^{3+}	—	—	—
$3d^8$	Ni^{2+}	84,420	61,110	–12,278

$^aB_{43} = \sqrt{10/7}\ B_{40}$

29.5 Bibliography and References

1. V. A. Antonov and P. A. Arsenev, *Spectroscopic Properties of Single Crystals of Rare Earth Titanates*, Phys. Status Solidi (a)**35** (1976), K169.

2. P.A.M. Berdowski and G. Blasse, *Luminescence and Energy Transfer in a Highly Symmetrical System: $Eu_2Ti_2O_7$*, J. Solid State Chem. **62** (1986), 317.

3. G. Campet, J. Claverie, and P. Salvador, *Influence of Cr^{3+} Doping on Photoelectronic Process of $La_2Ti_2O_7$ Pyrochlor and $La_{2/3}TiO_3$ Perovskite Anodes*, J. Phys. Chem. Solids **44** (1983), 925.

4. S. Fraga, K.M.S. Saxena, and J. Karwowski, *Handbook of Atomic Data*, Elsevier, New York (1976).

5. O. Knop, F. Brisse, and L. Castelliz, *Pyrochlores: V.—Thermoanalytic, X-ray, Neutron, Infrared, and Dielectric Studies of $A_2Ti_2O_7$*, Can. J. Chem. **47** (1969), 971.

6. K. A. Kuvshinova, M. L. Meil'men, A. G. Smagin, R. Yu. Abdulsaribov, V. A. Antonov, P. A. Arsen'ev, and L. B. Pasternak, *Structure of Active Centers in Manganese-Doped Single Crystals of Yttrium Titanate $Y_2Ti_2O_7$*, Sov. Phys. Crystallogr. **28** (1983), 198.

7. L. G. Mamsurova, K. S. Pigal'ski, K. K. Pukhov, and L. G. Scherbakova, *Elastic Properties and Acoustic Losses in Rare-Earth Compounds with the Pyrochlore Structure*, Sov. Phys. Solid State **30** (1988), 1349.

8. N. N. Mel'nik, L. M. Tsapenko, V. I. Larchev, and G. G. Skerotskaya, *High-Pressure Phases of Some Titanates of Rare-Earth Elements*, Izv. Akad. Nauk. SSSR, Neorg. Mater. **26** (1990), 805.

9. N. V. Porotnikov and G. Bazuev, *Vibrational Spectra of Samariam Titanate ($Sm_2Ti_2O_7$) and Europium Titanate ($Eu_2Ti_2O_7$) with Layered Perovskite Structure*, Zh. Neorg. Khim. **34** (1989), 250.

10. P. C. Schmidt, A. Weiss, and T. P. Das, *Effect of Crystal Fields and Self-Consistency on Dipole and Quadrupole Polarizabilities of Closed-Shell Ions*, Phys. Rev. **B19** (1979), 5525.

11. G. A. Teterin, I. S. Gainutdinov, I. M. Minaev, and L. V. Sadkovskaya, *Optical Properties and Inter-Particle Interaction Energy of $Ln_2Ti_2O_7$ Compounds*, Russ., J. Phys. Chem. **60** (1986), 1532.

12. M. T. Vandenborre and E. Husson, *Comparison of the Force Field in Various Pyrochlore Families. I. The $A_2B_2O_7$ Oxides*, J. Solid State Chem. **50** (1983), 362.

13. M. T. Vandenborre, E. Husson, J. P. Chatry, and D. Michel, *Rare-Earth Titanates and Stannates of Pyrochlore Structure; Vibrational Spectra and Force Fields*, J. Raman Spectrosc. **14** (1983), 63.

14. N. A. Zakhorov, V. S. Krikorov, E. F. Kustov, and S. Yu. Stefanovich, *New Non-linear Crystals in the $A_2B_2O_7$ Series*, Phys. Status Solidi **(a)50** (1978), K 13. (A = La, Ce, Pr, Nd, Sm, Gd, Y).

15. N. A. Zakharov, S. Yu. Stefanovich, V. S. Krikorov, and E. F. Kustov, *Syntheses of Several $A_2B_2O_7$ Crystals and Second-Harmonic Generation*, Sov. Tech. Phys. Lett. **4** (1978), 256.

16. N. A. Zakharov, S. Yu. Stefanovich, E. F. Kustov, and Yu. N. Venevtsev, *Growth of Crystals of Ferroelectrics $Ln_2Ti_2O_7$ (Ln = La – Nd) with Layer-Type Structures*, Krist. Tech. **15** (1980), 29.

30. $Y_2Ti_2O_7$

30.1 Crystallographic Data on $Y_2Ti_2O_7$ for Two Choices of Ion Sites

30.1.1 Cubic O_h^7 ($Fd3m$), 227 (second setting), Z = 8 (Knop et al, 1969)

Ion	Site	Symmetry	x^a	y	z	q	α (Å^3)^b
Y	16(c)	D_{3d}	0	0	0	3	0.870
Ti	16(d)	D_{3d}	1/2	1/2	1/2	4	0.22^c
O_1	8(a)	T_d	1/8	1/8	1/8	−2	1.34
O_2	48(f)	C_{2v}	0.4201	1/8	1/8	−2	1.349
X	8(b)	T_d	3/8	3/8	3/8	0	0

^{a}X-ray data: a = 10.095 Å (Knop et al, 1969).
^{b}Schmidt et al (1979).
^{c}Fraga et al (1976).

30.1.2 Cubic O_h^7 ($Fd3m$), 227 (second setting), Z = 8 (Becker, 1970)

Ion	Site	Symmetry	x^a	y	z	q	α (Å^3)^b
Y	16(d)	D_{3d}	1/2	1/2	1/2	3	0.870
Ti	16(c)	D_{3d}	0	0	0	4	0.22^c
O_1	8(b)	T_d	3/8	3/8	3/8	−2	1.349
O_2	48(f)	C_{2v}	−0.0788	1/8	1/8	−2	1.349
X	8(a)	T_d	1/8	1/8	1/8	0	0

^{a}X-ray data: a = 10.0896 Å (Becker, 1970).
^{b}Schmidt et al (1979).
^{c}Fraga et al (1976).

30.2 Crystal-Field Components, A_{nm}

The crystal-field components, A_{nm} (cm^{-1}/Å^n), are calculated through $n = 6$ because of the possibility of rare-earth ions occupying these sites.

30.2.1 For Y ion in 16c (D_{3d}) site of Knop et al (rotated so that z-axis is parallel to (111) crystallographic axis)

A_{nm}	Monopole	Self-induced	Dipole	Total
A_{20}	8,240	−1129	364	7476
A_{40}	11,645	−3145	109	8610
A_{43}	3,006	−731	−1447	829
A_{60}	1,954	−972	−104	878
A_{63}	−851	268	260	−322
A_{66}	1,182	−389	−206	586

30.2.2 **For Ti ion in 16d (D_{3d}) site of Knop et al (rotated so that z-axis is parallel to (111) crystallographic axis)**

A_{nm}	Monopole	Self-induced	Dipole	Total
A_{20}	−19,051	1887	14,909	−2,255
A_{40}	−14,726	5340	−5,417	−1,4802
A_{43}	22,131	−8636	2,985	16,480
A_{60}	4,242	−3071	293	1,463
A_{63}	−651	364	1,355	1,068
A_{66}	2,602	−1879	−48	675

30.2.3 **For vacancy site, X, 8b (T_d) of Knop et al**

$A_{nm}{}^a$	Monopole	Self-induced	Dipole	Total
A_{32}	−39,176	592	−790	38,978
A_{40}	34,880	−6419	6321	34,782
A_{60}	−708	−769	458	−1,019
A_{72}	−1,152	−66	0	−1,219

$^a A_{44} = \sqrt{5/14}\ A_{40}, A_{64} = -\sqrt{7/2}\ A_{60}, A_{76} = \sqrt{11/13}\ A_{72}$

30.2.4 **For Y ion in 16d (D_{3d}) site of Becker (rotated so that z-axis is parallel to (111) crystallographic axis)**

A_{nm}	Monopole	Self-induced	Dipole	Total
A_{20}	8,288	−1156	388	7519
A_{40}	11,686	−3152	84	8617
A_{43}	2,885	−705	−1368	812
A_{60}	1,952	−975	−92	885
A_{63}	−831	259	247	−325
A_{63}	1,167	−381	−197	590

30.2.5 **For Ti ion 16c (D_{3d}) site of Becker (rotated so that z-axis is parallel to (111) crystallographic axis)**

A_{nm}	Monopole	Self-induced	Dipole	Total
A_{20}	−17,718	1801	14,730	−1,188
A_{40}	−15,256	5588	−5,256	−14,924
A_{43}	22,458	−8832	2,833	16,458
A_{60}	4,282	−3128	221	1,375
A_{63}	−528	271	1,370	1,113
A_{66}	2,607	−1898	−58	651

30.2.6 For vacancy site, X, $8a$ (T_d) of Becker

A_{nm}	Monopole	Self-induced	Dipole	Total
A_{32}	39,189	597	−757	39,028
A_{40}	35,555	−6746	6306	35,115
A_{60}	−668	−812	462	−1,019
A_{72}	−1,157	−67	0	−1,224

$${}^a A_{44} = -\sqrt{5/14}\, A_{40}, A_{64} = \sqrt{7/2}\, A_{60}, A_{76} = -\sqrt{11/13}\, A_{72}$$

30.3 Experimental Parameters (cm⁻¹) (Antonov and Arsenev, 1976)

Ion	nd^N	$F^{(2)}$	$F^{(4)}$	$B_{40}{}^a$
V^{3+}	$3d^2$	63,217	41,378	−23,590
Cr^{3+}	$3d^3$	59,444	38,909	−25,970
Mn^{3+}	$3d^4$	—	—	−28,168
Co^{3+}	$3d^6$	56,630	34,020	−23,730
Mn^{4+}	$3d^3$	—	—	−31,822
Co^{2+}	$3d^7$	65,835	43,092	−8,330
Ni^{2+}	$3d^8$	84,420	61,110	−12,180

$${}^a B_{43} = \sqrt{10/7}\, B_{40}$$

30.4 Bibliography and References

1. V. A. Antonov and P. A. Arsenev, *Spectroscopic Properties of Single Crystals of Rare Earth Titanates*, Phys. Status Solidi (a)**35** (1976), K169.

2. V. A. Antonov, P. A. Arsenev, and D. S. Petrova, *Spectroscopic Properties of the Nd^{3+} Ion in $Y_2Ti_2O_7$ and $Gd_2Ti_2O_7$ Monocrystals*, Phys. Status Solidi (a)**41** (1977), K127.

3. W.-J. Becker, *Elektronen-Spin-Resonanz und optische Untersuchungen an Cr-Dotierten Y-Ti-Pyrochlor-Einkristallen $Y_2Ti_2O_7$*, Z. Naturforsch. **25a** (1970), 642. Also W.-J. Becker and G. Will, *Röntgen und Neutronenbeugungsuntersuchungen an $Y_2Ti_2O_7$*, Z. Kristallogr. **131** (1970), 278.

4. S. Fraga, K.M.S. Saxena, and J. Karwowski, *Handbook of Atomic Data*, Elsevier, New York (1976).

5. D. Goldschmidt, *Electronic Transport in $Y_2Ti_2O_7$: Structural Aspects of Small Polaron Formation*, Cryst. Latt. Def. Amorph. Mat. **15** (1987), 203.

6. O. Knop, F. Brisse, L. Castelliz, and O. Sutarno, *Determination of the Crystal Structure of Erbium Titanate, $Er_2Ti_2O_7$, by X-ray and Neutron Diffraction*, Can. J. Chem. **43** (1965), 2812. Also, O. Knop, F. Brisse,

and L. Castelliz, *Pyrochlores: V.—Thermoanalytic, X-ray, Neutron, Infrared, and Dielectric Studies of A$_2$Ti$_2$O$_7$ Titanates*, Can. J. Chem. **47** (1969), 971. Also see R.W.G. Wyckoff, *Crystal Structures*, vol 3, Interscience, New York (1968), p 441.

7. K. A. Kuvshinova, M. L. Meil'man, A. G. Smagin, R. Yu. Abdulsaribov, V. A. Antonov, P. A. Arsen'ev, and L. B. Pasternak, *Structure of Active Centers in Manganese-Doped Single Crystals of Yttrium Titanate Y$_2$Ti$_2$O$_7$*, Sov. Phys. Crystallogr. **28** (1983), 198.

8. P. C. Schmidt, A. Weiss, and T. P. Das, *Effect of Crystal Fields and Self-Consistency on Dipole and Quadrupole Polarizabilities of Closed-Shell Ions*, Phys. Rev. **B19** (1979), 5525.

31. K₂ReCl₆

31.1 Crystallographic Data on K_2ReCl_6

31.1.1 Cubic O_h^5 (Fm3m), 225, Z = 4

Ion	Site	Symmetry	x	y	z	q	α (Å³)
Re	$4a$	O_h	0	0	0	4	0.70^a
K	$8c$	T_d	1/4	1/4	1/4	1	0.827^b
Cl	$24e$	C_{4v}	x	0	0	−1	2.694^b

[a] Fraga et a (1976).
[b] Schmidt et al (1979).

31.1.2 X-ray data of K_2ReCl_6

a (Å)	X_{Cl}	Ref
9.861	0.240	12
9.7953	0.24037	11
9.840	0.2391	8

31.2 Crystal-Field Components, A_{nm} (cm⁻¹/Åⁿ), for Re (O_h) Site

$A_{nm}{}^a$	Monopole	Self-induced	Dipole	Total	Ref
A_{40}	5793	−4601	9,735	10,927	12
A_{40}	5959	−4823	10,129	11,264	8
A_{40}	5947	−4795	10,085	11,238	11

[a] $A_{40} = \sqrt{5/14}\ A_{40}$.

31.3 Experimental Parameters (cm⁻¹)

Ion	nd^N	$F^{(2)}$	$F^{(4)}$	ζ	B_{40}	Ref
Re⁴⁺	$5d^3$	31,891	20,928	2328	90,564	5

31.4 Bibliography and References

1. M. Bettinelli and C. D. Flint, *Magnon Sidebands and Cooperative Absorptions in K_2ReCl_6 and Cs_2ReCl_6*, J. Phys. **C21** (1988), 5499.

2. M. Bettinelli, L. DiSipio, G. Valle, C. Aschieri, and O. Ingletto, *Crystal Structures of Three Substituted Ammonium Hexachlororhenates (IV)*, Z. Kristallogr. **188** (1989), 155.

3. P. B. Dorain, *Magnetic and Optical Properties of Transition Metal Ions in Single Crystals*, Aerospace Laboratories Report, ARL-73-0139 (October 1973), NTIS AD 769870.

4. P. B. Dorain, *The Spectra of Re^{4+} in Cubic Crystals,* in *Transition Metal Chemistry*, vol 4, R. L. Carlin, ed., Marcel Dekker, New York (1968), p 1.

5. J. C. Eisenstein, *Magnetic Properties and Optical Absorption Spectrum of K$_2$ReCl$_6$*, J. Chem. Phys. **34** (1961), 1628. (This reference gives a large number of constants used in Eisenstein's calculation. We have taken his reported calculation and best fit it to obtain the parameters in sect. 31.3 above.)

6. S. Fraga, K.M.S. Saxena, and J. Karwowski, *Handbook of Atomic Data*, Elsevier, New York (1976).

7. H. D. Grundy and I. D. Brown, *A Refinement of the Crystal Structures of K$_2$ReCl$_6$, K$_2$ReBr$_6$ and K$_2$PtBr$_6$*, Can. J. Chem. **48** (1970), 1151.

8. V. J. Minkiewicz, G. Shirane, B. C. Frazer, R. G. Wheeler, and P. B. Dorain, *Neutron Diffraction Study of Magnetic Ordering in K$_2$IrCl$_6$, K$_2$ReBr$_6$ and K$_2$ReCl$_6$*, J. Phys. Chem. Solids **29** (1968), 881.

9. P. C. Schmidt, A. Weiss, and T. P. Das, *Effect of Crystal Fields and Self-Consistency on Dipole and Quadrupole Polarizabilities of Closed-Shell Ions*, Phys. Rev. **B19** (1979), 5525.

10. S.K.D. Strubinger, I-Wen Sun, W. E. Cleland, Jr., and C. L. Hussey, *Electrochemical and Spectroscopic Studies of Rhenium (IV) Monomeric and Dimeric Chloride Complexes in the Basic Aluminum-1-Methyl-3-Ethylimidazolium Chloride Room-Temperature Molten Salt*, Inorg. Chem. **29** (1990), 4246.

11. H. Takazawa, S. Ohba, and Y. Saito, *Electron-Density Distribution in Crystals of K$_2$[MCl$_6$] (M = Re, Os, Pt) and K$_2$[PtCl$_4$] at 120K*, Acta Crystallogr. **B46** (1990), 166.

12. R.W.G. Wyckoff, *Crystal Structures*, vol 3, Interscience, New York (1968), p 339.

13. R. K. Yoo, B. A. Kozikowski, S. C. Lee, and T. A. Keiderling, *Visible Region Absorption and Excitation Spectroscopy of K$_2$ReCl$_6$ and Various ReCl$_2^6$ Containing A$_2$MCl$_6$ Host Crystals*, Chem. Phys. **117** (1987), 255.

32. A_2BF_6 (A = K, Rb, Cs; B = Ge, Zr)

32.1   Crystallographic Data on A_2BF_6

Hexagonal D_{3d}^3 ($P\bar{3}m1$),164, $Z = 1$

Ion	Site	Symmetry	x	y	z	q	α (Å^3)
B	1(a)	D_{3d}	0	0	0	4	α_B
A	2(d)	C_{3v}	1/3	2/3	z	1	α_A
F	6(i)	C_s	x	$-x$	z	-1	0.731

32.2   X-Ray Data

A	B	a (Å)[a]	c (Å)	z_A	x_F	z_F	α_A (Å^3)[b]	α_B (Å^3)[c]
K	Ge	5.62	4.65	0.70	0.148	0.220	0.827	0.12
Rb	Ge	5.82	4.79	0.695	0.144	0.213	1.383	0.12
Cs	Zr	6.41	5.01	0.692	0.16	0.198	2.492	0.48
Rb	Zr	6.16	4.82	0.691	0.167	0.206	1.383	0.48

[a]Wyckoff (1968).

[b]Schmidt et al (1979).

[c]Fraga et al (1976).

32.3   Crystal-Field Components, A_{nm} (cm^{-1}/Å^n)

32.3.1   For Ge (D_{3d}) site of K_2GeF_6

A_{nm}	Monopole	Self-induced	Dipole	Total
A_{20}	3,371	−52.7	−2,040	1,278
A_{40}	−15,618	8,667	−18,481	−25,432
A_{43}	19,236	−10,270	23,136	32,103

32.3.2   For Ge (D_{3d}) site of Rb_2GeF_6

A_{nm}	Monopole	Self-induced	Dipole	Total
A_{20}	2,187	89.8	−2,280	−2.6
A_{40}	−15,138	8310	−17,768	−24,595
A_{43}	18,831	−9957	22,422	31,296

32.3.3   For Zr (D_{3d}) site of Cs_2ZrF_6

A_{nm}	Monopole	Self-induced	Dipole	Total
A_{20}	−10,846	1090	−4014	−13,771
A_{40}	−5,354	1942	−4819	−8,231
A_{43}	9,810	−3407	8329	14,732

32.3.4 For Zr (D_{3d}) site of Rb$_2$ZrF$_6$

A_{nm}	Monopole	Self-induced	Dipole	Total
A_{20}	−11,003	1047	−3653	−13,608
A_{40}	−5,288	1889	−4748	−8,147
A_{43}	9,738	−3347	8208	14,598

32.4 Bibliography and References

1. F. Averdunk and R. Hoppe, *On Single Crystal Syntheses of Semimetal Complex Fluorides: Without Solvents Li$_2$[GeF$_2$] and Na$_2$[GeF$_2$]*, Z. Anorg. Allg. Chem. **582** (1990), 111.

2. F. Averdunk and R. Hoppe, *Cubic Rb$_2$GeF$_6$ (With Observations About Rb$_3$GeF$_7$, Tl$_3$GeF$_7$ and Tl$_2$GeF$_6$)*, J. Fluorine Chem. **47** (1990), 481.

3. M. Bettinelli, L. DiSipio, G. Ingletto, and C. Razzetti, *Polarized Raman Spectra of the Trigonal Crystal K$_2$ReF$_6$*, Inorg. Chim. Acta **133** (1987), 7.

4. A. M. Black and C. D. Flint, *Splitting of the Γ_{8g} ($^4A_{2g}$) Ground State in Cs$_2$ReF$_6$*, J. Mol. Spectrosc. **70** (1978), 481.

5. A. M. Black and C. D. Flint, *Luminescence Spectra and Relaxation Processes of ReCl$_6^{2-}$ in Cubic Crystals*, J. Chem. Soc. Faraday Trans. 2 **73** (1976), 877.

6. A. M. Black and C. D. Flint, *Excitation and Luminescence Spectra of Dipotassium Hexafluoromanganate (IV)*, J. Chem. Soc. Dalton Trans. (1974), 977.

7. G. R. Clark and D. R. Russell, *Potassium Hexafluororhenate (IV)*, Acta Crystallogr. **B34** (1978), 894.

8. R. L. Davidovich and T. A. Kaidalova, *Ammonium Hexafluorostannate and Hexafluoroplumbate*, Russ. J. Inorg. Chem. **16** (1971), 1354.

9. S. Fraga, K.M.S. Saxena, and J. Karwowski, *Handbook of Atomic Data*, Elsevier, New York (1976).

10. F. Hanic, *The Crystal Chemistry of Complex Fluorides of General Formula A$_2$MF$_6$, The Refinement of the Structure (NH$_4$)$_2$SiF$_6$*, Chem. Zvesti **20** (1966), 738.

11. J. A. LoMenzo, S. Strobridge, H. H. Patterson, and H. Engstrom, *Electronic Absorption Spectra of the 5d^3 Hexafluororhenate (IV) Ion*, J. Mol. Spectrosc. **66** (1977), 150.

12. J. A. Martinez, M. A. Caracoche, P. C. Rivas, M. T. Dova, A. M. Rodriquez, and A. R. Lopez Garcia, *Phase Transitions in Rb$_2$ZrF$_6$ and Rb$_2$HfF$_6$ Using the Time-Differential Perturbed-Angular-Correlation Technique*, Phys. Rev. **B35** (1987), 5244.

A_2BF_6

13. P. C. Schmidt, A. Weiss, and T. P. Das, *Effect of Crystal Fields and Self-Consistency on Dipole and Quadrupole Polarizabilities of Closed-Shell Ions*, Phys. Rev. **B19** (1979), 5525.

14. R.W.G. Wyckoff, *Crystal Structures*, vol 3, Interscience, New York (1968), p 349.

15. Y. Yoshioka, N. Nakamura, and H. Chihara, *The ^{19}F Solid-State High-Resolution Nuclear Magnetic Resonance Study of K_2SiF_6, K_2GeF_6, and K_2SnF_6*, Bull. Chem. Soc. Jpn. **61** (1988), 3037.

33. Rb₂SnCl₆

33.1 Crystallographic Data on Rb₂SnCl₆

Cubic O_h^5 (Fm3m), 225, Z = 4

Ion	Site	Symmetry	x^a	y	z	q	$\alpha\,(\text{Å}^3)$
Sn	4(a)	O_h	0	0	0	4	0.37^b
Rb	8(c)	T_d	1/4	1/4	1/4	1	1.383^c
Cl	24(e)	C_{4v}	x	0	0	−1	2.694^c

^{a}X-ray data: a = 10.118 Å, x = 0.240 (Wyckoff, 1968).
^{b}Fraga et al (1976).
^{c}Schmidt et al (1979).

33.2 Crystal-Field Components, A_{nm} (cm⁻¹/Åⁿ), for Sn (O_h) Site

$A_{nm}{}^a$	Monopole	Self-induced	Dipole	Total
A_{40}	5094	−3739	8166	9521

$^aA_{44} = \sqrt{5/14}\ A_{40}$

33.3 Experimental Parameters (cm⁻¹)

Ion	nd^N	$F^{(2)}$	$F^{(4)}$	ζ	B_{40}	Ref
Os⁴⁺	$5d^4$	28,549	17,365	2606	40,198	13

$^aB_{44} = \sqrt{5/14}\ B_{40}$

33.4 Bibliography and References

1. A. Black and C. D. Flint, *Luminescence Spectra and Relaxation Processes of ReCl₆²⁻ in Cubic Crystals*, J. Chem. Soc. Faraday Trans. 2 **73** (1977), 877.

2. A. Black and C. D. Flint, *Jahn-Teller Effect in the $G_8(^2T_{2g}, t^3{}_{2g})$ State of ReBr₆²⁻*, J. Chem. Soc. Faraday Trans. 2 **71** (1975), 1871.

3. J. Degen, H. Kupka, and H.-H. Schmidtke, *Temperature-Dependent Luminescence Spectra of [ReCl₆]²⁻ Doped in K₂PtCl₆-Type Crystals*, Chem. Phys. **117** (1987), 163.

4. P. B. Dorain, *Magnetic and Optical Properties of Transition Metal Ions in Single Crystals*, Aerospace Laboratories Report, ARL-73-0139 (October 1973), NTIS AD 769870.

5. P. B. Dorain, *The Spectra of Re⁴⁺ in Cubic Crystals*, in *Transition Metal Chemistry*, vol 4, R. L. Carlin, ed., Marcel Dekker, New York (1968), p 1.

6. C. D. Flint and A. G. Paulusz, *High Resolution Infrared and Visible Luminescence Spectra of ReCl$_6^{2-}$ and ReBr$_6^{2-}$ in Cubic Crystals*, Mol. Phys. **43** (1981), 321.

7. C. D. Flint and A. G. Paulusz, *Infrared Luminescence Spectra of Hexachloroiridate (IV) and Hexabromoiridate (IV) in Cubic Crystals*, Inorg. Chem. **20** (1981), 1768.

8. C. D. Flint and A. G. Paulusz, *Infrared Luminescence of the ReBr$_6^{2-}$ Ion*, Chem. Phys. Lett. **62** (1979), 259.

9. S. Fraga, K.M.S. Saxena, and J. Karwowski, *Handbook of Atomic Data*, Elsevier, New York (1976).

10. J. Gilchrist, *Low Temperature Dielectric Relaxations in Ammonium Hexachlorostannate and Some Other Antifluorites*, J. Phys. Chem. Solids **50** (1989), 857.

11. R. J. Haskell and J. C. Wright, *Host Precipitates for Trace Analysis of Heavy Metal Ions by Laser Excited Fluorescence*, J. Lumin. **36** (1987), 331.

12. P. C. Schmidt, A. Weiss, and T. P. Das, *Effect of Crystal Fields and Self-Consistency on Dipole and Quadrupole Polarizabilities of Closed-Shell Ions*, Phys. Rev. **B19** (1979), 5525.

13. H.-H. Schmidtke and D. Strand, *The Emission Spectrum of OsCl$_6^{2-}$ Doped in Various Cubic Host Lattices*, Inorg. Chim. Acta **62** (1982), 153.

14. T. Schonherr, R. Wernicke, and H.-H. Schmidtke, *Luminescence Spectra and Electronic Levels of Osmium (IV) in Cubic and Quasi-Tetragonal Symmetry*, Spectrochim. Acta **38A** (1982), 679.

15. V. Waschk and J. Pelzl, *Raman Bands and Phase Transitions in $(A_xK_{1-x})_2SnCl_6$ Solid Solutions (A = Rb, NH$_4$)*, Raman Spectrosc. Proc. Int. Conf. 8th, J. Lascombe et al, ed. (1982), p 413.

16. R.W.G. Wyckoff, *Crystal Structures*, vol 3, Interscience, New York, (1968), p 339.

34. Cs_2TeCl_6

34.1 Crystallographic Data on Cs_2TeCl_6

34.1.1 Cubic O_h^5 (*Fm3m*), 225, Z = 4

Ion	Site	Symmetry	x^a	y	z	q	α (Å^3)
Te	4(a)	O_h	0	0	0	4	1.21[b]
Cs	8(c)	T_d	1/4	1/4	1/4	1	2.492[c]
Cl	24(e)	C_{4v}	x	0	0	−1	2.694[c]

[a]*X-ray data: a = 10.447 Å, x = 0.240 (Wyckoff, 1968).*
[b]*Fraga et al (1976).*
[c]*Schmidt et al (1979).*

34.1.2 X-ray data on similar compounds reported by Abriel (1987)

Compound	T (K)	α (Å)	x	Set
Rb_2TeBr_6	298	10.773	0.2496	1
Rb_2TeBr_6	160	10.713	0.2510	2
Co_2TeBr_6	298	10.934	0.2472	3
Cs_2TeBr_6	160	10.873	0.2486	4
Rb_2SnCl_6	293	10.137	0.2399	5
Rb_2SnCl_6	160	10.063	0.2412	6

For Rb, α = 1.383; Sn, α = 0.37; and Br, α = 3.263 Å^3.

34.2 Crystal-Field Components, A_{nm} (cm^{-1}/Å^n)

34.2.1 For Te (O_h) site (Wyckoff, 1968)

A_{nm}^a	Monopole	Self-induced	Dipole	Total
A_{40}	4341	−2885	6549	8005

[a]$A_{44} = \sqrt{5/14}\, A_{40}$

34.2.2 For 4(a) (O_h) site in compounds reported by Abriel (1987)

A_{nm}^a	Monopole	Self-induced	Dipole	Total	Set
A_{40}	3105	−2003	4596	5698	1
A_{40}	3113	−2003	4600	5710	2
A_{40}	3014	−1916	4431	5529	3
A_{40}	3020	−1915	4432	5538	4
A_{40}	5056	−3696	8085	9445	5
A_{40}	5114	−3753	8200	9562	6

[a]$A_{44} = \sqrt{5/14}\, A_{40}$

34.3   Bibliography and References

1. W. Abriel, *Vibronic Coupling and Dynamically Distorted Structures in Hexahalogenotellurates (IV): Low Temperature X-ray Diffraction (300–160K) and FTIR Spectroscopic (300–5K) Results*, Z. Naturforsch. **42b** (1987), 1273.

2. J. Degen and H.-H. Schmidtke, *Temperature-Dependent Luminescence Spectra and Lifetime Measurements of Octahedral Se(IV) and Te(IV) Hexahalogens Coordination Compounds*, Chem. Phys. **129** (1989), 483.

3. P. B. Dorain, *Magnetic and Optical Properties of Transition Metal Ions in Single Crystals*, Aerospace Laboratories Report, ARL-73-0139 (October 1973), NTIS AD 769870.

4. C. D. Flint, *Luminescence Spectra and Relaxation Processes of $ReCl_6^{2-}$ in Cubic Crystals*, J. Chem. Soc. Faraday Trans. 2 **74** (1978), 767.

5. C. D. Flint and P. Lang, *Infrared and Visible Luminescence of TcX_6^{2-} in Cubic Crystals*, J. Lumin. **24/25** (1981), 301.

6. C. D. Flint and A. G. Paulusz, *High Resolution Infrared and Visible Luminescence Spectra of $ReCl_6^{2-}$ and $ReBr_6^{2-}$ in Cubic Crystals*, Mol. Phys. **43** (1981), 321.

7. S. Fraga, K.M.S. Saxena, and J. Karwowski, *Handbook of Atomic Data*, Elsevier, New York (1976).

8. P. C. Schmidt, A. Weiss, and T. P. Das, *Effect of Crystal Fields and Self-Consistency on Dipole and Quadrupole Polarizabilities of Closed-Shell Ions*, Phys. Rev. **B19** (1979), 5525.

9. H.-H. Schmidtke and D. Strand, *The Emission Spectrum of $OsCl_6^{2-}$ Doped in Various Cubic Host Lattices*, Inorg. Chim. Acta **62** (1982), 153.

10. R.W.G. Wyckoff, *Crystal Structures*, vol 3, Interscience, New York (1968), p 339.

35. K_2ReF_6

35.1 Crystallographic Data on K_2ReF_6

Hexagonal D_{3d}^3 ($P\bar{3}m1$), 164, $Z = 1$

Ion	Site	Symmetry	x^a	y	z	q	α (Å^3)
Re	1(a)	D_{3d}	0	0	0	4	0.70^b
K	2(d)	C_{3v}	1/3	2/3	0.3007	1	0.827^c
F	6(i)	C_s	0.1617	–0.167	0.2276	–1	0.731^c

^{a}X-ray data: $a = 5.879$ Å, $c = 4.611$ Å (Clark and Russell, 1978).
^{b}Fraga et al (1976).
^{c}Schmidt et al (1979).

35.2 Crystal-Field Components, A_{nm} (cm^{-1}/Å^n), for Re (D_{3d}) Site

A_{nm}	Monopole	Self-induced	Dipole	Total
A_{20}	–4,231	592.3	–2,741	–6,380
A_{40}	–8,445	3421	–7,791	–12,815
A_{43}	–12,038	4732	–11,291	–18,597

35.3 Experimental Parameters (cm^{-1})

Ion	nd^N	$F^{(2)}$	$F^{(4)}$	ζ	B_{20}	B_{40}	B_{43}	Ref
Re^{4+}	$5d^3$	41,076	25,409	2612	–7136	–33,993	44,925	2

35.4 Bibliography and References

1. M. Bettinelli, L. DiSipio, G. Ingletto, and C. Flint, *Vibrational Dispersion in Ground and Excited States: The $\Gamma_7(^2T_{2g}) \leftrightarrow \Gamma_8(^4A_{2g})$ Absorption and Luminescence Spectra of Crystalline K_2ReF_6 at 4K*, Chem. Phys. Lett. **138** (1987), 361.

2. M. Bettinelli, L. Di Sipio, G. Ingletto, A. Montenero, and C. D. Flint, *Polarized Electronic Absorption Spectra of the Trigonal Crystal K_2ReF_6*, Mol. Phys. **56** (1985), 1033.

3. G. R. Clark and D. R. Russell, *Potassium Hexafluororhenate (IV)*, Acta Crystallogr. (**b**)**34** (1978), 894.

4. P. B. Dorain, *The Spectra of Re^{4+} in Cubic Fields*, in *Transition Metal Chemistry*, vol 4, R. L. Carlin, ed., Marcel Dekker, New York (1968), p 1.

5. S. Fraga, K.M.S. Saxena, and J. Karwowski, *Handbook of Atomic Data*, Elsevier, New York (1976).

6. P. C. Schmidt, A. Weiss, and T. P. Das, *Effect of Crystal Fields and Self-Consistency on Dipole and Quadrupole Polarizabilities of Closed-Shell Ions*, Phys. Rev. **B19** (1979), 5525.

36. Cs_2ZrCl_6

36.1 Crystallographic Data on Cs_2ZrCl_6

Cubic O_h^5 (Fm3m), 225, Z = 4

Ion	Site	Symmetry	x^a	y	z	q	α (Å^3)
Zr	4(a)	O_h	0	0	0	4	0.48^b
Cs	8(c)	T_d	1/4	1/4	1/4	1	2.492^c
Cl	24(e)	C_{4v}	x	0	0	−1	2.694^c

^{a}X-ray data: a = 10.407 Å, x = 0.235 (Wyckoff, 1968).
^{b}Fraga et al (1976).
^{c}Schmidt et al (1979).

36.2 Crystal-Field Components, A_{nm} (cm^{-1}/Å^n), for Zr (O_h) Site

$A_{nm}^{\ a}$	Monopole	Self-induced	Dipole	Total
A_{40}	4884	−3524	7742	9102

$^a A_{44} = \sqrt{5/14}\ A_{40}$

36.3 Experimental Parameters (cm^{-1})

Ion	nd^N	$F^{(2)}$	$F^{(4)}$	ζ	$B_{40}{}^a$	Ref
Ru^{4+}	$4d^4$	48,888	17,173	1044	39,732	13
Re^{4+}	$5d^3$	28,749	22,906	2392	63,729	7
Os^{4+}	$5d^4$	45,381	16,329	2416	47,229	6

$^a B_{44} = \sqrt{5/14}\ B_{40}$

36.4 Bibliography and References

1. J. F. Ackerman, *Preparation and Luminescence of Some [K$_2$PtCl$_6$] Materials*, Mater. Res. Bull. **19** (1984), 783.

2. J. C. Collingwood, P. N. Schatz, and P. J. McCarthy, *Absorption and Magnetic Circular Dichroism Spectra of Ru^{4+} in Cs$_2$ZrCl$_6$ and Cs$_2$SnBr$_6$*, Mol. Phys. **30** (1975), 269.

3. H. Donker, W. Van Schaik, W.M.A. Smit, and G. Blasse, *On the Luminescence of Selenium (IV) in A$_2$ZrCl$_6$ (A = Cs, Rb)*, Chem. Phys. Lett. **158** (1989), 509.

4. H. Donker, W.M.A. Smit, and G. Blasse, *On the Luminescence of Te^{4+} in A$_2$ZrCl$_6$ (A = Cs, Rb) and A$_2$SnCl$_6$ (A = Cs, Rb, K)*, J. Phys. Chem. Solids **50** (1989), 603.

5. P. B. Dorain, *Magnetic and Optical Properties of Transition Metal Ions in Single Crystals*, Aerospace Laboratories Report, ARL-73-0139 (October 1973), NTIS AD 769870.

6. P. B. Dorain, H. H. Patterson, and P. C. Jordan, *Optical Spectra of Os^{4+} in Single Cubic Crystals at 4.2 °K*, J. Chem. Phys. **49** (1968), 3845.

7. P. B. Dorain and R. G. Wheeler, *Optical Spectrum of Re^{4+} in Single Crystals of K$_2$PtCl$_6$ and Cs$_2$ZrCl$_6$ at 4.2 °K*, J. Chem. Phys. **45** (1966), 1172.

8. P.J.H. Drummen, H. Donker, W.M.A. Smit, and G. Blasse, *Jahn-Teller Distortion in the Excited State of Tellurium (IV) in Co$_2$MCl$_6$ (M = Zr, Sn)*, Chem. Phys. Lett. **144** (1988), 460.

9. C. D. Flint and P. A. Tanner, *Vibronic Spectra of U^{4+} in Octahedral Crystal Fields IV. Absorption Spectra and Crystal Field Calculations*, Molec. Phys. **61** (1987), 389.

10. S. Fraga, K.M.S. Saxena, and J. Karwowski, *Handbook of Atomic Data*, Elsevier, New York (1976).

11. S. M. Khan, H. H. Patterson, and H. Engstrom, *Multiple State Luminescence for the d^4 OsCl$_6^{2-}$ Impurity Ion in K$_2$PtCl$_6$ and Cs$_2$ZrCl$_6$ Cubic Crystals*, Mol. Phys. **35** (1978), 1623.

12. B. A. Kozikowski and T. A. Keiderling, *Intraconfigurational Absorption and Magnetic Circular Dichroism Spectra of Os^{4+} in Cs$_2$ZrCl$_6$ and in Cs$_2$ZrBr$_6$*, Mol. Phys. **40** (1980), 477.

13. H. Patterson and P. Dorain, *Optical Spectra of Ru^{4+} in Single Crystals of K$_2$PtCl$_6$ and Cs$_2$ZrCl$_6$ at 4.2°K*, J. Chem. Phys. **52** (1970), 849.

14. A. R. Reinberg and S. G. Parker, *Sharp-Line Luminescence of Re^{4+} in Cubic Single Crystals of Cs$_2$ZrCl$_6$ and Cs$_2$HfCl$_6$*, Phys. Rev. **B1** (1970), 2085.

15. P. C. Schmidt, A. Weiss, and T. P. Das, *Effect of Crystal Fields and Self-Consistency on Dipole and Quadrupole Polarizabilities of Closed-Shell Ions*, Phys. Rev. **B19** (1979), 5525.

16. H.-H. Schmidtke and D. Strand, *The Emission Spectrum of OsCl$_6^{2-}$ Doped in Various Cubic Host Lattices*, Inorg. Chim. Acta **62** (1982), 153.

17. R.W.G. Wyckoff, *Crystal Structures*, vol 3, Interscience, New York (1968), p 339.

18. R. K. Yoo and T. A. Keiderling, *Intraconfigurational Absorption Spectroscopy of IrCl$_6^{2-}$ and IrBr$_6^{2-}$ in A$_2$MX$_6$-Type Host Crystals*, Chem. Phys. **108** (1986), 317.

19. R. K. Yoo, S. C. Lee, B. A. Kozikowski, and T. A. Keiderling,
     *Intraconfigurational Absorption Spectroscopy of $ReCl_6^{2-}$ in Various
     A_2MCl_6 Host Crystals*, Chem. Phys. **117** (1987), 237.

20. R. K. Yoo and T. A. Keiderling, *Intraconfigurational Absorption and
     Two-Photon Excitation Spectra of $PtCl_6^{2-}$-Containing Crystals*, J. Phys.
     Chem. **94** (1990), 8048.

37. GaAs

37.1 Crystallographic Data on GaAs

Cubic $T_d^2 (F\bar{4}3m)$, 216, $Z = 4$

Ion	Site	Symmetry	x	y	z	q	α (Å³)[a]
Ga	4a	T_d	0	0	0	3	4.435
As	4c	T_d	1/4	1/4	1/4	–3	3.786
X	4b	T_d	1/2	1/2	1/2	—	—

$a = 5.6537$ Å (Wyckoff, 1963).

[a]Pandy et al (1977).

37.2 Crystal-Field Components, A_{nm} (cm^{-1}/Å^n), for X (T_d) Site

A_{nm}[a]	Monopole	Self-induced	Total
A_{32}	19,129	–4,174	14,956
A_{40}	–12,451	372.6	–12,078
A_{60}	400	–532	–132
A_{72}	–349	329	–20

[a]$A_{44} = \sqrt{5/14}\ A_{40}$, $A_{64} = \sqrt{7/2}\ A_{60}$, $A_{76} = \sqrt{11/13}\ A_{72}$

37.3 Crystal Fields for Ga Site

37.3.1 Crystal-field components, A_{nm} (cm^{-1}/Å^n), for Ga (T_d) site

A_{nm}[a]	Monopole	Self-induced	Total
A_{32}	19,129	–4,174	14,956
A_{40}	–5,544	2,238	–3,306
A_{60}	638.8	–420.4	218.4
A_{72}	–348.6	328.9	–19.63

[a]$A_{44} = \sqrt{5/14}\ A_{40}$, $A_{64} = \sqrt{7/2}\ A_{60}$, $A_{76} = \sqrt{11/13}\ A_{72}$

37.3.2 Crystal-field parameters, B_{nm} (cm^{-1}), for triply ionized transition-metal ions[a]

X^{3+}	$3d^N$	$F^{(2)}$	$F^{(4)}$	B_{40}[b]	ζ[c]
Ti	d^1	—	—	−16,802	144
V	d^2	31,218	28,343	−12,436	200
Cr	d^3	33,316	30,226	−9,739	269
Mn	d^4	35,325	32,023	−7,829	353
Fe	d^5	37,401	33,889	−6,314	455
Co	d^6	38,947	35,231	−5,487	568
Ni	d^7	40,667	36,748	−4,704	705
Cu	d^8	42,418	38,296	−4,045	865
Zn	d^9	—	—	−3,495	1050

[a]*C. A. Morrison and D. E. Wortman (1990).*

[b]$B_{44} = \sqrt{5/14}\ B_{40}$

[c]*Obtained by fitting the experimental data of Clerjaud et al (1985). Parameters obtained for the other ions are interpolated.*

37.4 Bibliography and References

1. G. A. Allen, *The Activation Energies of Chromium, Iron and Nickel in Gallium Arsenide*, J. Phys. **D1** (1968), 593.

2. F. G. Anderson and F. S. Hamm, *Detecting the Dynamic Aspect of the Jahn-Teller Coupling for the V^{3+} Defect in GaAs*, Mater. Sci. Forum **38–41** (1989), 305.

3. D. G. Andrianov, N. I. Suchkova, A. S. Sarel'ev, E. P. Rashevskaya, and M. A. Filippov, *Magnetic and Optical Properties of Ni^{3+} and Co^{2+} Ion of $3d^7$ Configuration in Gallium Arsenide*, Sov. Phys. Semicond. **11** (1977), 426.

4. G. Aszodi and U. Kaufmann, *Zeeman Spectroscopy of the Vanadium Luminescence in GaP and GaAs*, Phys. Rev. **B32** (1985), 7108.

5. F. Bantien, G. Hofmann, K. Reimann, and J. Weber, *Manganese Levels in GaAs Under Hydrostatic Pressure, in AlGaAs, and in GaAs/AlGaAs Quantum Well—A Comparative Study*, Mater. Science Forum **38–41** (1989), 1403.

6. G. A. Baraff, *Stress Splitting of the Zero-Phonon Line of the (As_{Ga} − As_{Si}) Defect Pair in GaAs: Significance for the Identity of EL2*, Phys. Rev. **B40** (1984), 1030.

7. C. A. Bates and D. Brugel, *The Ground State of Trigonal $Cr^{2+}(II)$:GaAs: Refinements in the Model from Low-Frequency Phonon Scattering*, Semicond. Sci. Technol. **2** (1987), 494.

8. C. A. Bates, M. Darcha, J. Handley, A. Varson, and A. M. Vassum, *A Study of Isolated Substitutional Cr^{2+} in GaAs by Thermally Detected EPR*, Semicond. Sci. Technol. **3** (1988), 172.

9. C. A. Bates and J. L. Dunn, *The Jahn-Teller Effect and Zero Phonon Line Intensities*, Defect and Diffusion Forum **62/63** (1989), 27.

10. C. A. Bates, J. L. Dunn, and W. Ulrici, *Strong Jahn-Teller Effects in Excited $^3T_1$ State of V^{3+} Ions in III-V Materials*, J. Phys. Condens. Matter **2** (1990), 607.

11. S. W. Biernacki and B. Clerjaud, *Optical Spectra of Ti, V and Co in III-V and II-VI Crystals*, Acta Phys. Pol. **A73** (1988), 251.

12. J. E. Bisberg, F. P. Dabkowski, and A. K. Chin, *Cubic Zircon as a Species Permeable Coating for Zinc Diffusion in Gallium Arsenide*, Appl. Phys. Lett. **53** (1988), 18.

13. R. Braunstein, S. M. Eetemadi, and R. K. Kim, *Deep Level Derivative Spectroscopy of Semiconductors by Wavelength Modulation Techniques*, SPIE **524**, Spectroscopic Characterization Techniques for Semiconductor Technology II (1985), 51.

14. R. Braunstein, R. K. Kim, D. Matthews, and M. Braunstein, *Derivative Absorption Spectroscopy of GaAs:Cr*, Physica **117B,118B** (1983), 163.

15. D. Brugel and C. A. Bates, *The Trigonal Cr^{2+} (II) Centre in GaAs: Refinements in the ^5E Excited State*, Semicond. Sci. Technol. **2** (1987), 501.

16. M. R. Burd and R. Braunstein, *Deep Levels in Semi-insulating Liquid Encapsulated Czochralski-Grown GaAs*, J. Phys. Chem. Solids **49** (1988), 731.

17. M. J. Caldos, A. Fazzio, and A. Zunger, *A Universal Trend in the Binding Energies of Deep Impurities in Semiconductors*, Appl. Phys. Lett. **45** (1984), 671.

18. M. J. Caldos, S. K. Figueiredo, and A. Gazzio, *Theoretical Investigation of the Electrical and Optical Activity of Vanadium in GaAs*, Phys. Rev. **B33** (1986), 7102.

19. B. Clerjaud, *Transiton-Metal Impurities in III-V Compounds*, J. Phys. **C18** (1985), 3615.

20. B. Clerjaud, D. Cote, F. Gendron, M. Krause, and W. Ulrici, *Isotopic Effects in GaAs:Ni*, Mater. Sci. Forum **38–41** (1989), 775.

21. B. Clerjaud, C. Naud, B. Deveaud, B. Lambert, B. Plot, G. Bremond, C. Benjiddou, G. Guillot, and A. Nousilhat, *The Acceptor Levels of Vanadium in III-V Compounds*, J. Appl. Phys. **58** (1985), 4207.

22. E. S. Demidov, *Impurity States of Iron Group Ions in Gallium Arsenide and Silicon*, Sov. Phys. Solid State **19** (1977), 100.

23. E. S. Demidov, A. A. Ezhevskii, and V. V. Karzanov, *Excited States of the Fe^{3+} Ion in Gallium Arsenide and Phosphide*, Sov. Phys. Semicond. **17** (983), 412.

24. S. M. Eetemadi and R. Braunstein, *Wavelength Modulation Absorption Spectroscopy of Deep Levels in Semi-Insulating GaAs*, J. Appl. Phys. **58** (1985), 2217.

25. M. En-Nagadi, A. Vasson, A. M. Vasson, C. A. Bates, and A. F. Labadz, *A V^{2+} Center in GaAs Studied by Thermally Detected EPR*, J. Phys. **C21** (1988), 1137.

26. L. K. Ermakov, V. F. Masterov, and R. E. Samorukov, *Electron Structure of Transition Elements in Indium Phosphide*, Sov. Phys. Semicond. **18** (1984), 1304.

27. G. F. Fengand and R. Zallern, *Optical Properties of Ion-Impurities in GaAs: The Observation of Finite-Size Effects in GaAs Microcrystals*, Phys. Rev. **B40** (1989), 1064.

28. A. Gorger, B. K. Meyer, J. M. Spaeth, and A. M. Hennel, *Identification of Spin, Charge States, and Optical Transitions of Vanadium Impurities in GaAs*, Semicond. Sci. Technol. **3** (1988), 832.

29. D. P. Halliday, C. A. Payling, M. K. Saker, M. S. Skolnick, W. Ulrici, and L. Eaves, *Zeeman Spectroscopy on Ti-Doped GaAs and GaP*, Semicond. Sci. Technol. **2** (1987), 679.

30. A. M. Hennel, C. D. Brandt, K. Y. Yo, and L. M. Pawlowicz, *Properties of Titanium in GaAs and InP*, Mater. Sci. Forum **10–12** (1986), 645.

31. A. M. Hennel, C. D. Brandt, Y. T. Wu, T. Bryskiewicz, K. Y. Yo, J. Lagewski, and H. C. Gatos, *Absorption Spectra of Ti-Doped GaAs*, Phys. Rev. **B33** (1986), 7353.

32. A. M. Hennel, C. D. Brandt, K. Y. Yo, J. Lagowski, and H. C. Gatos, *Optical and Electronic Properties of Vanadium in Gallium Arsenide*, J. Appl. Phys. **62** (1987), 163.

33. V. A. Kasatkin, A. A. Lavrent'ev, and P. A. Rodnyi, *Kinetics of the Electro-Luminescence of Yttrium Ions in Indium Phosphide*, Sov. Phys. Semicond. **19** (1985), 221.

34. U. Kaufmann, H. Ennen, J. Scheinder, R. Worner, J. Weber, and F. Kohl, *Spectroscopic Study of Vanadium in GaP and GaAs*, Phys. Rev. **B25** (1982), 5598.

35. V. I. Kirillov, A. I. Spirin, and V. S. Postnikov, *Influence of Uniaxial Compressing on the ESR Spectra of Transition Metal Ions in III-V Compounds,* Sov. Phys. Solid State **28** (1986), 1796.

36. S. Lee, J. D. Dow, and O. F. Sankay, *Theory of Charge States Splitting of Deep Levels,* Phys. Rev. **B31** (1985), 3910.

37. A. Louati, T. Benyattou, P. Roura, G. Bremond, G. Gaillot, and R. Coquille, *Photo-Luminescence Study of Fe^{2+} and Local Atomic Arrangements in $In_{1-x}Ga_xP$ Alloys,* Mater. Sci. Forum **38–41** (1989), 935.

38. R. M. Macfarlane and J. C. Vial, *Saturation Spectroscopy and Fluorescence of $ZnO:Co^{2+}$,* Mater. Sci. Forum **10–12** (1986), 845.

39. D.T.E. Marple, *Index of Refraction of GaAs,* J. Appl. Phys. **35** (1964), 1941.

40. V. A. Morozova and V. V. Ostoborodova, *Electro-Modulation Spectra of High Resistance Chromium-Doped p-Type and n-Type Gallium Arsenide,* Vestn. Mosk. Univ. Ser. Fiz. Astron. **24** (1983), 67 (in Russian).

41. C. A. Morrison and D. E. Wortman, *Analysis of the Internal Optical Spectra of Triply Ionized Transition-Metal Ions in III-V Semiconductors,* Proc. AAAS Annual Meeting, New Orleans, LA, USA (15–20 February 1990).

42. H. Nakagoma, K. Uwai, and K. Takahei, *Extremely Sharp Erbium-Related Intra-4f-Shell Photo-Luminescence of Erbium-Doped GaAs Grown by Metal-Organic Chemical Vapor Desposition,* Appl. Phys. Lett. **53** (1988), 1726.

43. T. Nishino, T. Yanagida, and Y. Hamakawa, *Impurity Electro Absorption of GaAs,* Jpn. J. Appl. Phys. **11** (1972), 1221.

44. R. N. Pandy, T. P. Sharma, and B. Dayal, *Electronic Polarization of Ions in Group III-V Compounds,* J. Phys. Chem. Solids **38** (1977), 329.

45. C. A. Payling, D. P. Halliday, D. G. Hayes, M. K. Saker, M. S. Skolnick, W. Ulrici, and L. Eaves, *Optical Properties and Zeeman Spectroscopy of Ti-Doped GaP and GaAs,* Proc. Fourteenth International Symposium on Gallium Arsenide and Related Compounds, Heraklion, Crete (28 September–1 October 1987), Inst. of Physics Conference series number 91, chapter 2 (1988), p 125.

46. P. Roura, T. Benyattou, G. Guillot, R. Moncorge, and W. Ulrici, *A Study of the $^3T_2 \to {}^3A_2$ Photoluminescence of Ti^{2+} in GaP,* Semicond. Sci. Technol. **4** (1989), 943.

47. E. Sigmund and C. A. Bates, *A Theoretical Investigation of Deep-Level Impurities in Semiconductors: Chromium and Manganese in GaAs*, Phil. Mag. **B56** (1987), 611.

48. J.A.L. Simpson, C. A. Bates, J. Barrau, M. Brousseau, and V. Thomas, *A Study of the Cr^{2+}-Te Complex in GaAs*, Semicond. Sci. Technol. **3** (1988), 178.

49. H. J. Schulz, *Transition Metal Impurities in Compound Semiconductors: Experimental Situation*, Mater. Chem. Phys. **16** (1987), 373.

50. H. J. Schultz and M. Theide, *Excitation Processes of GaAs:$V^{3+}(d^2)$*, J. Phys. **C21** (1988), L1033.

51. V. Thomas, J. Barrau, G. Armelles, B. Deveaud, and B. Lambert, *The Luminescence at 0.844 eV in GaAs:Cr—A Zeeman Spectroscopy*, Mater. Sci. Forum **10-12** (1986), 675.

52. H. Tokumoto and T. Ishiguro, *Photoacoustic Study of Chromium in GaAs*, Jpn. J. Appl. Phys. **22** (1983), 202.

53. W. T. Tsang and R. A. Logan, *Observation of Enhanced Single Longitudinal Mode Operation in 1.5 μm GaInAsP Erbium-Doped Semiconductor Injection Laser*, Appl. Phys. Lett. **49** (1986), 1689.

54. W. Ulrici, *Photo-Induced Optical Absorption of Cr^{4+} in Semi-insulating GaAs:Cr*, Phys. Status Solidi **(b)131** (1985), 707.

55. W. Ulrici, L. Eaves, K. Friedland, D. P. Halliday, and J. Kreilbl, *Vanadium in GaAs and GaP*, 14th International Conference on Defects in Semiconductors (18–22 August 1986), vol 10–12 (1986), p 639.

56. W. Ulrici, L. Eaves, K. Friedland, D. P. Halliday, K. J. Nash, and M. S. Skolnick, J. Phys. **C19** (1986), L525.

57. W. Ulrici, K. Friedland, L. Eaves, and D. P. Halliday, *Optical and Electrical Properties of Vanadium-Doped GaAs*, Phys. Status Solidi **(b)131** (1985), 719.

58. V. V. Ushakov, A. A. Gippius, V. A. Dravin, and A. V. Spitsyn, *Luminescence of a Rare-Rarth (Erbium) Impurity in Gallium Arsenide and Phosphide*, Sov. Phys. Semicond. **16** (1982), 723.

59. J. P. Vander Ziel, M. G. Ober, and R. A. Logan, *Single Mode Operation of Er-Doped 1.5 μm InGaAsP Laser*, Appl. Phys. Lett. **50** (1987), 1313.

60.   J. Wagner, *Optical Spectroscopy of Impurity Levels in GaAs*, Phys. Scr. **T29** (1989), 167.

61.   S. Watanabe and H. Kamimura, *First-Principles Calculation of Multiplet Structures of Transition Metal Deep Impurities in II-VI and III-V Semiconductors*, Mater. Sci. Eng. **B3** (1989), 313.

62.   S. Watanabe and H. Kamimura, *Multiplet Structure of Transition Metal Deep Impurities in ZnS*, J. Phys. Soc. Jpn. **56** (1987), 1078.

63.   S. Watanabe and H. Kamimura, *Multiplet Structures of Transition Metal Deep Impurities in GaAs*, J. Phys. **C20** (1987), 4145.

64.   R.W.G. Wyckoff, *Crystal Structures*, vol 1, Interscience, New York (1963), p 110.

65.   A. Wysmolck, Z. Liro, and A. M. Hennel, *High Resolution Measurements of the $^3A_2 \rightarrow \,^3T_2$ Absorption Spectrum in V-Doped GaAs*, Mater. Science Forum **38–42** (1989), 827.

66.   A. Wysmdek and A. M. Hennel, *Photoionization Spectra of Ion Doped InP and GaAs*, Acta. Phys. Pol. **A77** (1990), 67.

67.   M. Zigone, H. R. Buisson, and G. Martinez, *Study of Cr^{2+} Luminescence in GaAs as a Function of Hydrostatic Pressure*, Mater. Sci. Forum **10–12** (1986), 663.

38. ZrSiO$_4$

38.1 Crystallographic Data on ZrSiO$_4$

38.1.1 Tetragonal D_{4h}^{19} ($4_1/amd$), 141 (second setting)

Ion	Site	Symmetry	x	y	z	q	α (Å^3)
Zr	$4a$	D_{2d}	0	3/4	1/8	4	0.48[a]
Si	$4b$	D_{2d}	0	1/4	3/8	4	0.03[a]
O	$16h$	C_s	0	y	z	-2	1.349[b]

[a]Fraga et al (1976).
[b]Schmidt et al (1979).

38.1.2 X-ray data on ZrSiO$_4$

a (Å)	c (Å)	y	z	Ref
6.6164	6.0150	0.067	0.198	a
6.607	5.982	0.0661	0.1953	b

[a]Wyckoff (1968).
[b]Robinson et al (1971).

38.2 Crystal Fields for Zr (D_{2d}) Site

38.2.1 Crystal-field components, A_{nm} (cm^{-1}/Å^n), for Zr (D_{2d}) site (data of Wyckoff, 1968)

A_{nm}	Monopole	Self-induced	Dipole	Total
A_{20}	$-12,765$	349.4	33,644	21,229
A_{32}	$-1,332$	449.2	958.9	76.55
A_{40}	910.3	-1130	524.3	304.5
A_{44}	8,630	-2943	$-7,031$	$-1,344$
A_{52}	6,142	-2379	$-1,796$	1,967

38.2.2 Crystal-field components, A_{nm} (cm^{-1}/Å^n), for Zr (D_{2d}) site (data of Robinson et al, 1971)

A_{nm}	Monopole	Self-induced	Dipole	Total
A_{20}	$-13,840$	388.8	34,843	21,392
A_{32}	-522.2	332.7	754.3	564.8
A_{40}	826.4	-1175	369.0	20.10
A_{44}	9,024	-3114	$-7,316$	$-1,406$
A_{52}	6,357	-2514	$-1,703$	2,142

38.2.3 Theoretical crystal-field parameters, B_{nm} (cm^{-1}), for quadruply ionized 4d transition metals[a]

X^{4+}	nd^N	$F^{(2)}$[b]	$F^{(4)}$	ζ	B_{20}[c]	B_{40}	B_{44}
Nb	$4d^1$	0	0	742	−11,125	1419	13,454
Mo	$4d^2$	68,068	45,102	914	−9,599	989	9,364
Tc	$4d^3$	71,843	47,615	1,105	−8,985	910	8,624
Ru	$4d^4$	75,495	50,038	1,318	−7,607	617	5,849
Rh	$4d^5$	79,206	52,512	1,556	−7,049	526	4,985
Pd	$4d^6$	82,196	54,434	1,819	−6,473	439	4,162
Ag	$4d^7$	85,389	56,519	2,114	−5,941	365	3,456
Cd	$4d^8$	88,615	58,631	2,441	−5,361	291	2,762
La	$4d^9$	0	0	2,805	−4,851	218	2,071

[a]*Morrison (1990).*

[b]*$F^{(2)}$, $F^{(4)}$, and ζ are Hartree-Fock values (cm^{-1}).*

[c]*The crystal-field parameters are $B_{nm} = <r^n> A_{nm}$, $<r^n>$ are Hartree-Fock, and A_{nm} are from 38.2.1.*

38.2.4 Theoretical crystal-field parameters, B_{nm} (cm^{-1}), for quadruply ionized 5d transition metals[a]

X^{4+}	nd^N	$F^{(2)}$[b]	$F^{(4)}$	ζ	B_{20}[c]	B_{40}	B_{44}
Ta	$5d^1$	0	0	2792	−12,539	1686	15,984
W	$5d^2$	64,633	43,252	3257	−11,140	1243	11,783
Re	$5d^3$	67,496	45,208	3741	−10,645	1180	11,190
Os	$5d^4$	70,197	47,082	4259	−9,251	839.0	7,954
Ir	$5d^5$	72,980	49,004	4814	−8,733	741.1	7,026
Pt	$5d^6$	75,122	50,424	5404	−8,161	639.4	6,062
Au	$5d^7$	77,448	51,994	6045	−7,619	548.9	5,204
Hg	$5d^8$	79,808	53,587	6736	−7,021	455.2	4,315
Tl	$5d^9$	0	0	7486	−6,510	364.1	3,452

[a]*Morrison (1990).*

[b]*$F^{(2)}$, $F^{(4)}$, and ζ are Hartree-Fock values (cm^{-1}).*

[c]*The crystal-field parameters are $B_{nm} = <r^n> A_{nm}$, $<r^n>$ are Hartree-Fock, and A_{nm} are from 38.2.1.*

38.3 Bibliography and References

1. G. Bayer, *Thermal Expansion of ABO$_4$–Compounds with Zircon and Scheelite Structure*, J. Less-Common Metals **26** (1972), 255.

2. R. Caruba, A. Baumer, and P. Hartman, *Crystal Growth of Synthetic Zircon Round Natural Seeds*, J. Cryst. Growth **88** (1988), 297.

3. S. Di Gregario, M. Greenblatt, and J. H. Pifer, *ESR of Nb^{4+} in Zircon*, Phys. Status Solidi **(b)101** (1980), K149.

4. S. Di Gregario, M. Greenblatt, J. H. Pifer, and M. D. Sturge, *An ESR and Optical Study of V^{4+} in Zircon-Type Crystals*, J. Chem. Phys. **76** (1982), 2931.

5. S. Fraga, J. Karwowski, and K.M.S. Saxena, *Handbook of Atomic Data*, vol 5, Elsevier, New York (1976), p 319.

6. E. A. Harris, J. H. Mellor and S. Parke, *Electron Paramagnetic Resonance of Tetravalent Praseodymium in Zircon*, Phys. Status Solidi **(b)122** (1984), 757.

7. Hong Xiaoyu, Bai Gui-ru, and Zhao Min-guang, *The Study of the Optical and the EPR Spectra of V^{4+} in Zircon-Type Crystals*, J. Phys. Chem. Solids **46** (1985), 719.

8. G. F. Koster, J. O. Dimmock, R. G. Wheeler, and H. Statz, *Properties of the Thirty-Two Point Groups*, MIT Press, Cambridge, MA (1963).

9. K. Kusabu, T. Yagi, M. Kikuchi, and Y. Syono, *Structural Considerations on the Mechanism of the Shock-Induced Zircon-Scheelite Transition in ZrSiO$_4$*, J. Phys. Chem. Solids **47** (1986), 675.

10. K. B. Lyons, M. D. Sturge, and M. Greenblatt, *Low-Frequency Raman Spectrum of ZrSiO$_4$:V^{4+}: An Impurity-Induced Dynamical Distortion*, Phys. Rev. **B30** (1984). 2127.

11. C. A. Morrison, *Host Materials for $4d^N$ and $5d^N$ Transition-Metal Ions*, Harry Diamond Laboratories, HDL-TM-90-20 (December 1990).

12. D. J. Newman and Betty Ng, *Crystal-Field Superposition Model Analysis for Tetravalent Actinides*, J. Phys: Condens. Mater. **1** (1989), 1613.

13. I. S. Poirot, W. K. Kot, N. M. Edelstein, M. M. Abraham, C. B. Finch, and L. A. Boatner, *Optical Study and Analysis of Pu^{4+} in Single Crystals of ZrSiO$_4$*, Phys. Rev. **B39** (1989), 6388.

14. I. Poirot, W. Kot, G. Shallimoff, N. Edelstein, M. M. Abraham, C. B. Finch, and L. A. Boatner, *Optical and E.P.R. Investigations of Np^{4+} in Single Crystals of ZrSiO$_4$*, Phys. Rev. **B37** (1988), 3255.

15. K. Robinson, G. V. Gibbs, and P. H. Ribbe, *The Structure of Zircon: A Comparison with Garnet*, Amer. Mineral. **56** (1971), 782.

16. P. C. Schmidt, A. Weiss, and T. P. Das, *Effects of Crystal Fields and Self-Consistency on Dipole and Quadruple Polarizabilities of Closed-Shell Ions*, Phys. Rev. **B19** (1979), 5525.

17. L. M. Silich, E. M. Kurpan, N. M. Bobkora, A. A. Stepanchuk, and S. A. Gailevich, *Quantitative X-Ray Phase Analysis of Ceramic in the Al_2O_3-TiO_2 System with AlO_2 and ZrSiO$_4$ Added*, Neorg. Mater. **24** (1988), 1196.

18. R.W.G. Wyckoff, *Crystal Structures*, vol 4, Interscience, New York (1968), p 157.

39. HfGeO$_4$

39.1 Crystallographic Data on HfGeO$_4$

Tetragonal C_{4h}^6 ($I4_1/a$), 88 (first setting), $Z = 4$

Ion	Site	Symmetry	x	y	z	q	α (Å^3)
Hf	4b	S_4	0	0	1/2	4	0.57[a]
Ge	4a	S_4	0	0	0	4	0.12[a]
O	16f	C_1	x	y	z	−2	1.349[b]

[a]*Fraga et al (1976).*
[b]*Schmidt et al (1979).*

39.2 X-Ray Data on HfGeO4

a (Å)	c (Å)	x	y	z
4.849[a]	10.50	0.25	0.11	0.07
4.862[b]	10.497	0.2678	0.1739	0.0831

[a]*Wyckoff (1968).*
[b]*Ennaciri et al (1986).*

39.3 Crystal-Field Components, A_{nm} (cm^{-1}/Å^n), for Hf (S_4) Site of HfGeO$_4$ (Wyckoff, 1968)

A_{nm}	Monopole	Self-induced	Dipole	Total
A_{20}	6465	−662.8	−13,794	−7992
ReA_{32}	3527	−434.0	4,855	7948
ImA_{32}	512.3	−45.50	674.9	1141
A_{40}	−706.6	389.5	3,039	2721
ReA_{44}	−2525	755.7	139.9	−1629
ImA_{44}	−3920	961.5	2,154	−804.2
ReA_{52}	3026	−966.8	−904.1	1155
ImA_{52}	−3758	1289	2,735	267.2
$|A_{44}|$	4663	—	—	1817

39.4 Bibliography and References

1. G. Beyer, *Thermal Expansion of ABO$_4$ Compounds with Zircon and Scheelite Structures*, J. Less-Common Metals **26** (1972), 255.

2. A. Ennaciri, A. Kahn, and D. Michel, *Crystal Structures of HfGeO$_4$ and ThGeO$_4$ Germinates*, J. Less-Common Metals **124** (1986), 105.

3. S. Fraga, J. Karwowski, and K.M.S. Saxena, *Handbook of Atomic Data*, vol 5, Elsevier, New York (1976), p 319.

4.   C. A. Morrison, *Host Materials for 4d^N and 5d^N Transition-Metal Ions*, Harry Diamond Laboratories, HDL-TM-90-20 (December 1990).

5.   P. C. Schmidt, A. Weiss, and T. P. Das, *Effect of Crystal Fields and Self-Consistency on Dipole and Quadruple Polarizabilities of Closed-Shell Ions*, Phys. Rev. **B19** (1979), 5525.

6.   R.W.G. Wyckoff, *Crystal Structures*, vol 3, Interscience, New York (1968), p 21.

40. Li_2XTeO_6 (X = Zr, Hf)

40.1 Crystallographic Data on Li_2XTeO_6 (X = Zr, Hf)

Trigonal C_3^4 ($R3$), 146 (hexagonal setting), $Z = 1$

Ion	Site	Symmetry	x	y	z	q	α ($Å^3$)
Li_1	$3a$	C_3	0	0	z	1	0.0321^a
Li_2	$3a$	C_3	0	0	z	1	0.0321^a
X	$3a$	C_3	0	0	z	4	$a_x{}^b$
Te	$3a$	C_3	0	0	z	6	0.20^b
O_1	$9b$	C_1	x	y	z	-2	1.349^a
O_2	$9b$	C_1	x	y	z	-2	1.349^a

^{a}Schmidt et al (1979).
^{b}Fraga et al (1976).

40.2 X-Ray Data (Choisnet et al, 1988)

X	a (Å)	c (Å)	z_{Li1}	z_{Li2}	z_X	z_{Te}	x_{O1}	y_{O1}	z_{O1}
Zr	5.172	13.847	0.29	0.76	0.993	0.500	0.049	0.355	0.077
Hf	5.164	13.782	—	—	—	—	—	—	—

X	x_{O2}	y_{O2}	z_{O2}	α_x (Å^3)
Zr	0.652	0.962	0.576	0.48
Hf	—	—	—	0.57

40.3 Crystal-Field Components, A_{nm} (cm^{-1}/Å^n), for Zr Site (C_3) of Li_2ZrTeO_6

A_{nm}	Monopole	Self-induced	Dipole	Total
A_{10}	−16,240	—	−5318	−21,557
A_{20}	457.0	26.77	5159	5,643
A_{30}	2,319	−971.7	5941	7,289
ReA_{33}	1,582	−239.1	−5242	−3,899
ImA_{33}	−2,633	881.9	−4924	−6,675
A_{40}	−14,546	5506	467.2	−8,573
ReA_{43}	3,925	−1773	9713	11,865
ImA_{43}	17,820	−6293	−4997	6,530
A_{50}	1,239	−610.4	−828.5	−200.1
ReA_{53}	451.4	−38.55	−1981	−1,568
ImA_{53}	−1,805	1241	−958.1	−1,522
$|A_{43}|$	18,247	—	—	13,543

40.4 Crystal-Field Components, A_{nm} (cm^{-1}/Å^n), for Hf Site (C_3) of Li_2HfTeO_6

The position of the ions within the unit cell are those of Li_2ZrTeO_6.

A_{nm}	Monopole	Self-induced	Dipole	Total
A_{10}	68,052	—	9,798	77,850
A_{20}	251.1	55.40	7,418	7,724
A_{30}	2,387	–998.3	–28,186	–26,797
ReA_{33}	1,607	–243.4	1,478	2,842
ImA_{33}	–2,677	903.0	21,103	19,330
A_{40}	–14,696	5605	3,543	–5,548
ReA_{43}	3,972	–1,811	11,185	13,345
ImA_{43}	18,073	–6,432	–9,130	2,511
A_{50}	1,249	–617.3	4,789	5,421
ReA_{53}	453.8	–38.22	–1,007	–591.8
ImA_{53}	–1,834	1270	4,337	3,773
$|A_{43}|$	18,504	—	—	13,579

40.5 Bibliography and References

1. J. Choisnet, A. Rulmont, and P. Tarte, *Les Tellurates mixtes Li_2ZrTeO_6 et Li_2HfTeO_6: un nouveau phénomène d'ordre dans la famille corindou*, J. Solid State Chem. **75** (1988), 124.

2. S. Fraga, J. Karwowski, and K.M.S. Saxena, *Handbook of Atomic Data*, vol 5, Elsevier, New York (1976), p 319.

3. C. A. Morrison, *Host Materials for $4d^N$ and $5d^N$ Transition-Metal Ions*, Harry Diamond Laboratories, HDL-TM-90-20 (December 1990).

4. P. C. Schmidt, A. Weiss, and T. P. Das, *Effects of Crystal Fields and Self-Consistency on Dipole and Quadruple Polarizabilities of Closed-Shell Ions*, Phys. Rev. **B19** (1979), 5525.

41. Li_6BeZrF_{12}

41.1 Crystallographic Data on Li_6BeZrF_{12}

Tetragonal D_{4h}^{19} ($I41/amd$), 141 (second setting), $Z = 4$

Ion	Site	Symmetry	x^a	y	z	q	$\alpha\,(\text{Å}^3)^b$
Be	$4a$	D_{2d}	0	3/4	1/8	2	0.0125
Zr	$4b$	D_{2d}	0	1/4	3/8	4	0.480
Li_1	$8e$	C_{2v}	0	1/4	0.1034	1	0.0321
Li_2	$16f$	C_2	0.2303	0	0	1	0.0321
F_1	$16h$	C_s	0	0.5340	0.4207	−1	0.731
F_2	$16h$	C_s	0	0.0260	0.2903	−1	0.731
F_3	$16h$	C_s	0	−0.0568	0.0745	−1	0.731

^{a}X-ray data: $a = 6.570$, $c = 18.62$ (Wyckoff, 1968).

^{b}Schmidt et al (1979), except for Zr, which is from Fraga et al (1976).

41.2 Crystal-Field Components, A_{nm} $(\text{cm}^{-1}/\text{Å}^n)$, for Zr (D_{2d}) Site

A_{nm}	Monopole	Self-induced	Dipole	Total
A_{20}	1411	97.22	−6180	−4672
A_{32}	−3101	763.4	−4360	−6697
A_{40}	−4960	1762	−2039	−5237
A_{44}	5393	−1943	3955	7405
A_{52}	3606	−1754	2452	4304

41.3 Bibliography and References

1. S. Fraga, J. Karwowski, and K.M.S. Saxena, *Handbook of Atomic Data*, vol 5, Elsevier, New York (1976).

2. C. A. Morrison, *Host Materials for $4d^N$ and $5d^N$ Transition-Metal Ions*, Harry Diamond Laboratories, HDL-TM-90-20 (December 1990).

3. P. C. Schmidt, A. Weiss, and T. P. Das, *Effect of Crystal Fields and Self-Consistency on Dipole and Quadruple Polarizabilities of Closed-Shell Ions*, Phys. Rev. **B19** (1979), 5525.

4. R.W.G. Wyckoff, *Crystal Structures*, vol 4, Interscience, New York (1968), p 57.

42. ZrGeO₄

42.1 Crystallographic Data on ZrGeO₄

Tetragonal C_{4h}^6 ($I4_1/a$), 88 (first setting), $Z = 4$

Ion	Site	Symmetry	x^a	y	z	q	$\alpha\,(\text{Å}^3)^b$
Ge	4a	S_4	0	0	0	4	0.12^b
Zr	4b	S_4	0	0	1/2	4	0.48^b
O	16f	C_1	x	y	z	−2	1.349

[a]X-ray data: a = 4.8660, c = 10.55 (Å), x = 0.2664, y = 0.1726, z = 0.0822 (Ennaciri et al, 1984).

[b]Fraga et al (1976).

[c]Schmidt et al (1979).

42.2 Crystal-Field Components, A_{nm} (cm⁻¹/Åⁿ), for Zr (S_4) Site

A_{nm}	Monopole	Self-induced	Dipole	Total
A_{20}	5175	−202	−13,444	−8470
ReA_{32}	−2584	543	3,702	1661
ImA_{32}	5083	−1228	2,074	5928
A_{40}	−6023	1773	7,119	2870
ReA_{44}	−5668	1928	−154	−3894
ImA_{44}	−4887	1457	50	−3380
ReA_{52}	1982	−754	394	1622
ImA_{52}	−5705	2280	717	−2708
$|A_{44}|$	7484	—	—	5516

42.3 Crystal-Field Components, A_{nm} (cm⁻¹/Åⁿ), for Ge (S_4) Site

A_{nm}	Monopole	Self-induced	Dipole	Total
A_{20}	−16,423	2,648	−17,512	−31,288
ReA_{32}	17,821	−5,603	8,684	20,902
ImA_{32}	36,558	−12,128	21,800	46,235
A_{40}	−16,132	7,646	−3,565	−12,051
ReA_{44}	−10,322	5,533	−6,628	−11,417
ImA_{44}	12,115	−6,078	9,726	15,764
ReA_{52}	−2,489	1,770	−3,199	−3,918
ImA_{52}	−5,540	3,989	−6,510	−8,062
$|A_{44}|$	15,916	—	—	19,464

42.4   Bibliography and References

1.   G. Bayer, *Thermal Expansion of ABO$_4$ Compounds with Zircon and Scheelite Structures*, J. Less-Common Metals **26** (1972), 255.

2.   A. Ennaciri, D. Michel, M. Perez y Jorba, and J. Pannetier, *Neutron Diffraction Determination of the Structure of an Ordered Scheelite—Type: Zr$_3$GeO$_8$*, Mater. Res. Bull. **19** (1984), 793.

3.   S. Fraga, K.M.S. Saxena, and J. Karwowski, *Handbook of Atomic Data*, vol 5, Elsevier, New York (1976).

4.   C. A. Morrison, *Host Materials for 4d^N and 5d^N Transition-Metal Ions*, Harry Diamond Laboratories, HDL-TM-90-20 (December 1990).

5.   P. C. Schmidt, A. Weiss, and T. P. Das, *Effect of Crystal Fields and Self-Consistency on Dipole and Quadrupole Polarizabilities of Closed-Shell Ions*, Phys. Rev. **B19** (1979), 5525.

43. Zr₃GeO₈

43.1 Crystallographic Data on Zr_3GeO_8

Tetragonal D_{2d}^{11} ($\bar{I}42m$), 121 (first setting), $Z = 2$

Ion	Site	Symmetry	x^a	y	z	q	$\alpha\,(\text{Å}^3)^b$
Ge	2a	D_{2d}	0	0	0	4	0.12^b
Zr_1	2b	D_{2d}	0	0	1/2	4	0.48^b
Zr_2	4d	S_4	0	1/2	1/4	4	0.48^b
O_1	8i	C_s	0.2004	0.2004	0.3410	−2	1.349^c
O_2	8i	C_s	0.2170	0.2170	0.0904	−2	1.349^c

^{a}X-ray data: $a = 5.005$, $c = 10.550$ (Å) (Ennaciri et al, 1984).
^{b}Fraga et al (1976).
^{c}Schmidt et al (1979).

43.2 Crystal-Field Components, A_{nm} (cm⁻¹/Åⁿ), for Ge (D_{2d}) Site of Zr_3GeO_8

A_{nm}	Monopole	Self-induced	Dipole	Total
A_{20}	−9,412	1,462	−13,999	−21,949
A_{32}	37,705	−11,897	26,755	52,564
A_{40}	−17,256	7,914	−7,401	−16,742
A_{44}	13,709	−6,580	10,516	−17,644
A_{52}	−3,458	2,382	−5,316	−6,393

43.3 Crystal-Field Components, A_{nm} (cm⁻¹/Åⁿ), for Zr_1 (D_{2d}) Site of Zr_3GeO_8

A_{nm}	Monopole	Self-induced	Dipole	Total
A_{20}	10,659	−900	−7933	1827
A_{32}	300.1	38.3	−98.4	240.6
A_{40}	−7,875	2384	6546	1055
A_{44}	6,875	−2024	−91.5	4760
A_{52}	−5,795	2320	−289.0	−3764

43.4 Crystal-Field Components, A_{nm} (cm^{-1}/Å^n), for Zr_2 (S_4) Site of Zr_3GeO_8

A_{nm}	Monopole	Self-induced	Dipole	Total
A_{20}	−7,440	974	−6987	−13,453
ReA_{32}	−6,953	1611	129	−5,219
ImA_{32}	−12,700	3350	−5948	−15,298
A_{40}	−9,538	2874	3317	−3,347
ReA_{44}	−6,213	2229	−170	−4,154
ImA_{44}	6,389	−2090	2204	6,500
ReA_{52}	1,489	−645	1029	1,873
ImA_{52}	3,755	−1660	1473	3,568
$|A_{44}|$	8,910	—	—	7,714

43.5 Bibliography and References

1. A. Ennaciri, D. Michel, M. Perez y Jorba, and J. Pannetier, *Neutron Diffraction Determination of the Structure of an Ordered Scheelite—Type: Zr_3GeO_8*, Mater. Res. Bull. **19** (1984), 793.

2. S. Fraga, K.M.S. Saxena, and J. Karwowski, *Handbook of Atomic Data*, vol 5, Elsevier, New York (1976).

3. C. A. Morrison, *Host Materials for $4d^N$ and $5d^N$ Transition-Metal Ions*, Harry Diamond Laboratories, HDL-TM-90-20 (December 1990).

4. P. C. Schmidt, A. Weiss, and T. P. Das, *Effect of Crystal Fields and Self-Consistency on Dipole and Quadrupole Polarizabilities of Closed-Shell Ions*, Phys. Rev. **B19** (1979), 5525.

44. ThSiO$_4$

44.1 Crystallographic Data on ThSiO$_4$

Tetragonal D_{4h}^{19} ($I4_1/amd$), 141 (first setting), $Z = 4$

Ion	Site	Symmetry	x^a	y	z	q	α (Å^3)
Th	4a	D_{2d}	0	3/4	1/8	4	1.52[b]
Si	4b	D_{2d}	0	3/4	5/8	4	0.030[b]
O	16h	C_s	0	0.0732	0.2104	−2	1.349[c]

[a]X-ray data: a = 7.1328 Å, c = 6.3188 Å (Taylor et al, 1978).
[b]Fraga et al (1976).
[c]Schmidt et al (1979).

44.2 Crystal-Field Components, A_{nm} (cm^{-1}/Å^n), for Th Site (D_{2d}) of ThSiO$_4$

A_{nm}	Monopole	Self-induced	Dipole	Total
A_{20}	−6300	−141.4	26,500	20,058
A_{32}	−1653	193.5	1,666	206.2
A_{40}	18.45	−493.3	1,350	875.4
A_{44}	4906	−1316	−4,142	−552.9
A_{52}	3753	−1108	−1,116	1,529

44.3 Crystal-Field Components, A_{nm} (cm^{-1}/Å^n), for Si Site (D_{2d}) of ThSiO$_4$

A_{nm}	Monopole	Self-induced	Dipole	Total
A_{20}	7,634	−3,788	28,562	32,408
A_{32}	−64,599	24,940	−45,571	−85,230
A_{40}	−35,517	21,569	−31,618	−46,566
A_{44}	13,907	−9,471	10,306	14,741
A_{52}	−7,113	6,834	−14,295	−14,573

44.4   Bibliography and References

1.   S. Fraga, J. Karwowski, and K.M.S. Saxena, *Handbook of Atomic Data*, vol 5 (1976), p 319.

2.   C. A. Morrison, *Host Materials for 4d^N and 5d^N Transition-Metal Ions*, Harry Diamond Laboratories, HDL-TM-90-20 (December 1990).

3.   P. C. Schmidt, A. Weiss, and T. P. Das, *Effects of Crystal Fields and Self-Consistency on Dipole and Quadrupole Polarizabilities of Closed-Shell Ions*, Phys. Rev. **B19** (1979), 5525.

4.   M. Taylor and R. C. Ewing, *The Crystal Structures of the ThSiO$_4$ Polymorphs: Huttonite and Thorite*, Acta Crystallogr. **B34** (1978), 1074.

45. ThGeO₄

45.1 Crystallographic Data on ThGeO₄

45.1.1 Tetragonal D_{4h}^{19} ($I4_1/amd$), 141 (first setting), Z = 4

Ion	Site	Symmetry	x^a	y	z	q	α (Å³)
Th	4a	D_{2d}	0	0	0	4	1.52[b]
Ge	4b	D_{2d}	0	0	1/2	4	0.12[b]
O	16h	C_s	0	0.1803	0.3214	-2	1.349[c]

[a]X-ray data: a = 7.230 Å, c = 6.539 Å (Ennaciri et al, 1986).
[b]Fraga et al (1976).
[c]Schmidt et al (1979).

45.1.2 Tetragonal C_{4h}^6 ($I4_1/a$), 88 (first setting), Z = 4

Ion	Site	Symmetry	x^a	y	z	q	α (Å³)
Th	4b	S_4	0	0	1/2	4	1.52[b]
Ge	4b	S_4	0	0	0	4	0.12[b]
O	16f	C_1	0.2548	0.1493	0.0787	-2	1.348[c]

[a]X-ray data: a = 5.145 Å, c = 10.531 Å (Ennaciri et al, 1986).
[b]Fraga et al (1976).
[c]Schmidt et al (1979).

45.2 Crystal-Field Components, A_{nm} (cm⁻¹/Åⁿ), for Th Site (D_{2d}) of ThGeO₄

A_{nm}	Monopole	Self-induced	Dipole	Total
A_{20}	-8188	49.52	21,620	13,482
A_{32}	151.6	-68.33	171.6	254.8
A_{40}	549.2	-547.2	448.5	450.6
A_{44}	5368	-1397	-3,701	270.1
A_{52}	-3634	1074	829.8	-1,730

45.3 Crystal-Field Components, A_{nm} (cm⁻¹/Åⁿ), for Ge Site (D_{2d}) of ThGeO₄

A_{nm}	Monopole	Self-induced	Dipole	Total
A_{20}	15,677	-3,858	19,380	31,119
A_{32}	48,087	-15,117	27,015	59,985
A_{40}	-26,107	12,492	-17,436	-31,051
A_{44}	8,289	-4,775	5,642	9,156
A_{52}	7,228	-5.656	8,759	10,331

45.4 Crystal-Field Components, A_{nm} (cm^{-1}/Å^n), for Th Site (S_4) of ThGeO$_4$

A_{nm}	Monopole	Self-induced	Dipole	Total
A_{20}	3593	−382.9	−4985	−1775
ReA_{32}	1850	−150.2	3916	5616
ImA_{32}	−98.96	33.57	124.6	59.21
A_{40}	−3236	977.3	3822	1563
ReA_{44}	−3857	977.1	1069	−1811
ImA_{44}	−3787	888.8	810.3	−2088
ReA_{52}	1923	−599.4	138.0	1461
ImA_{52}	−3970	1326	1369	−1275
$|A_{44}|$	5405	—	—	2764

45.5 Crystal-Field Components, A_{nm} (cm^{-1}/Å^n), for Ge Site (S_4) of ThGeO$_4$

A_{nm}	Monopole	Self-induced	Dipole	Total
A_{20}	−21,877	3,538	−13,762	−32,101
ReA_{32}	22,842	−7,897	15,609	30,554
ImA_{32}	40,279	−13,830	23,195	49,644
A_{40}	−16,030	8,434	−7,823	−15,419
ReA_{44}	−9,474	5,340	−3,575	−7,709
ImA_{44}	16,075	−8,588	13,046	20,533
ReA_{52}	−3,667	2,846	−4,027	−4,849
ImA_{52}	−6,414	5,096	−5,729	−7,048

45.6 Bibliography and References

1. G. Bayer, *Thermal Expansion of ABO$_4$ Compounds with Zircon and Scheelite Structures*, J. Less-Common Metals **26** (1972), 255.

2. A. Ennaciri, A. Kahn, and D. Michel, *Crystal Structures of HfGeO$_4$ and ThGeO$_4$ Germanates*, J. Less-Common Metals **124** (1986), 105.

3. S. Fraga, J. Karwowski, and K. M. S. Saxena, *Handbook of Atomic Data*, vol 5, Elsevier, New York (1976), p 319.

4. C. A. Morrison, *Host Materials for 4d^N and 5d^N Transition-Metal Ions*, Harry Diamond Laboratories, HDL-TM-90-20 (December 1990).

5. P. C. Schmidt, A. Weiss, and T. P. Das, *Effects of Crystal Fields and Self-Consistency on Dipole and Quadrupole Polarizabilities of Closed-Shell Ions*, Phys. Rev. **B19** (1979), 5525.

46. Na₂TiSiO₅

46.1 Crystallographic Data on Na₂TiSiO₅

Tetragonal D_{4h}^7 ($P4/nmm$), 129, $Z = 2$

Ion	Site	Symmetry	x^a	y	z	q	α $(\text{Å}^3)^b$
Ti	$2c$	C_{4v}	1/2	0	0.9343	4	0.506
Na	$4e$	C_{2h}	1/4	1/4	1/2	1	0.147
Si	$2a$	D_{2d}	0	0	0	4	0.0165
O₁	$8i$	C_s	0	0.2071	0.1831	−2	1.349
O₂	$2c$	C_{4v}	1/2	0	−0.7338	−2	1.349

^{a}X-ray data: $a = 6.480$ Å, $c = 5.107$ Å (Nyman et al, 1978).
^{b}Schmidt et al (1979).

46.2 Crystal-Field Components, A_{nm} (cm⁻¹/Åⁿ), for Si (D_{2d}) Site in Na₂TiSiO₅

A_{nm}	Monopole	Self-induced	Dipole	Total
A_{20}	7,285	548.7	−34,349	−26,515
A_{32}	−64,975	25,258	−48,610	−88,327
A_{40}	−31,427	19,391	−20,545	−32,581
A_{44}	17,004	−12,063	21,554	26,494
A_{52}	326.3	−628.4	9,042	8,740

46.3 Crystal-Field Components, A_{nm} (cm⁻¹/Åⁿ), for Ti (C_{4v}) Site in Na₂TiSiO₅

A_{nm}	Monopole	Self-induced	Dipole	Total
A_{10}	15,041	0	114,675	129,716
A_{20}	18,952	−2,906	38,432	54,478
A_{30}	48,794	−13,854	36,992	71,932
A_{40}	18,057	−10,303	42,160	49,914
A_{44}	10,890	−4,552	766.4	7,105
A_{50}	4,284	−5,673	25,232	23,843
A_{52}	326.3	−628.4	9,042	8,740

46.4 Bibliography and References

1. Yu. K. Egorov Tismenko, M. A. Simonov, and N. V. Belov, *A Revised Crystal Structure for Synthetic Sodium Titanosilicate $Na(TiO)[SiO_4]$*, Sov. Phys. Dokl. **23** (1978), 289.

2. C. A. Morrison, *Possible Hosts for Quadruply Ionized $3d^N$ Transition Metal Ions: Na_2TiSiO_4, $Y_2SiBe_2O_7$, $Bi_4X_3O_{12}$, and $Bi_{12}XO_{20}$ (X = Si, Ge)*, Harry Diamond Laboratories, HDL-TM-91-1 (May 1991).

3. H. Nyman, M. O'Keeffe, and J. O. Borin, *Sodium Titanium Silicate, Na_2TiSiO_5*, Acta Crystallogr. **B34** (1978), 905.

4. P. C. Schmidt, A. Weiss, and T. P. Das, *Effect of Crystal Fields and Self-Consistency on Dipole and Quadrupole Polarizabilities of Closed-Shell Ions*, Phys. Rev. **B19** (1979), 5525.

47. $Y_2SiBe_2O_7$

47.1 Crystallographic Data on $Y_2SiBe_2O_7$

Tetragonal D_{3d}^3 $(P\bar{4}2_1m)$, 113, $Z = 2$

Ion	Site	Symmetry	x^a	y	z	q	$\alpha\,(\text{Å}^3)^b$
Y	$4e$	C_s	0.1595	0.6595	0.4873	3	0.870
Si	$2a$	S_4	0	0	0	4	0.0165^c
Be	$4e$	C_s	0.363	0.863	0.031	2	0.0125
O_1	$8f$	C_1	0.0823	0.1664	0.7928	-2	1.349
O_2	$4e$	C_2	0.3561	0.8561	0.7053	-2	1.349
O_3	$2c$	C_{2v}	0	1/2	0.8275	-2	1.349

^{a}X-ray data: $a = 7.283$ Å, $c = 4.755$ Å (Bartram, 1969).
^{b}Schmidt et al (1979).
^{c}Tessman et al (1953).

47.2 Crystal-Field Components, A_{nm} (cm^{-1}/Å^n), for Si Site (S_4) in $Y_2SiBe_2O_7$

A_{nm}	Monopole	Self-induced	Dipole	Total
A_{20}	10,029	-468.9	$-19,678$	$-10,118$
ReA_{32}	35,565	$-12,974$	26,656	49,247
ImA_{32}	$-46,373$	17,029	$-40,428$	$-69,772$
A_{40}	$-28,792$	16,971	$-23,527$	$-35,348$
ReA_{44}	$-4,073$	2,468	$-6,484$	$-8,089$
ImA_{44}	$-14,462$	9,193	$-17,375$	$-22,644$
ReA_{52}	902.0	-662.1	$-2,652$	$-2,412$
ImA_{52}	$-1,164$	866.0	3,394	3,096
$|A_{44}|$	15,025	—	—	24,045

47.3 Bibliography and References

1. S. F. Bartram, *Crystal Structure of Y$_2$SiBe$_2$O$_7$*, Acta Crystallogr. **B25** (1969), 791.

2. A. A. Ismatov, *Synthesis of Melilite-Like Minerals Containing Rare Earth Ions*, Dokl. Akad. Nauk Uzb. SSR **30** (1973), 50.

3. C. A. Morrison, N. Karayianis, and D. E. Wortman, *Rare-Earth Ion-Host Interactions: 10. Lanthanides in Y$_2$SiBe$_2$O$_7$*, Harry Diamond Laboratories, HDL-TR-1766 (August 1976).

4. P. C. Schmidt, A. Weiss, and T. P. Das, *Effect of Crystal Fields and Self-Consistency on Dipole and Quadruple Polarizabilities of Closed-Shell Ions*, Phys. Rev. **B19** (1979), 5525.

5. J. R. Tessman, A. H. Kahn, and W. Schockley, *Electronic Polarizabilities of Ions in Crystals*, Phys. Rev. **92** (1953), 890.

48. $Bi_4X_3O_{12}$ (X = Si, Ge)

48.1 Crystallographic Data on $Bi_4X_3O_{12}$

Cubic T_d^6 ($I\bar{4}3d$) 220, $Z = 4$

Ion	Site	Symmetry	x	y	z	q	α (Å^3)
Bi	16c	C_3	x	x	x	3	2.23[a]
X	12a	S_4	0	1/4	3/8	4	$\alpha_x{}^a$
O	48e	C_1	x	y	z	–4	1.349[b]

[a]Fraga et al (1976), $\alpha_{Si} = 0.03$ Å^3, $\alpha_{Ge} = 0.12$ Å^3.
[b]Schmidt et al (1979).

48.2 X-Ray Data on $Bi_4X_3O_{12}$

X	a (Å)	x_{Bi}	x_O	y_O	z_O
Si	10.300[a]	0.0857	0.0607	0.1335	0.2875
Ge	10.513[b]	0.0876	0.0689	0.1277	0.2875

[a]Wyckoff (1968).
[b]Fisher and Waldner (1982).

48.3 Crystal-Field Components, A_{nm} (cm^{-1}/Å^n), for Si Site (S_4) of $Bi_4Si_3O_{12}$

A_{nm}	Monopole	Self-induced	Dipole	Total
A_{20}	–4,423	1,103	11,851	8,531
ReA_{32}	39,156	–15,088	35,345	59,414
ImA_{32}	55,809	–21,662	43,774	77,921
A_{40}	–29,886	19,317	–36,749	–47,317
ReA_{44}	–7,631	4,551	–3,239	–6,319
ImA_{44}	18,425	–12,613	15,697	21,509
ReA_{52}	–2,043	1,506	2,727	2,190
ImA_{52}	–2,628	2,150	3,917	3,439
$|A_{44}|$	19,943	—	—	22,418

48.4 Crystal-Field Components, A_{nm} (cm^{-1}/Å^n), for Ge Site (S_4) of $Bi_4Ge_3O_{12}$

A_{nm}	Monopole	Self-induced	Dipole	Total
A_{20}	−9,149	1,563	7,098	−488.4
ReA_{32}	26,496	−8,445	21,045	39,096
ImA_{32}	43,745	−14,017	29,123	58,851
A_{40}	−19,352	10,242	−22,771	−31,881
ReA_{44}	−7,790	3,872	−2,562	−6,480
ImA_{44}	13,318	−7,527	8,938	14,729
ReA_{52}	−2,366	1,523	1,117	274.0
ImA_{52}	−3,549	2,507	1,981	938.8
$|A_{44}|$	19,943	—	—	22,418

48.5 Bibliography and References

1. F. Allegretti, D. Borgia, R. Riva, F. DeNotaristefani, and S. Pizzini, *Growth of BGO Single Crystals Using a Directional Solidification Technique*, J. Cryst. Growth **94** (1989), 373.

2. T. M. Bochkova, E. G. Valyashko, V. A. Smirnov, and S. A. Flerova, *Optical Spectra of $Bi_4Ge_3O_{12}$-Nd and $Bi_4Si_3O_{12}$-Nd Crystals*, J. Appl. Spectrosc. (USSR) **30** (1979), 102.

3. D. P. Bortfeld and H. Meier, *Refractive Indices and Electro-Optic Coefficients of the Eulitites $Bi_4Ge_3O_{12}$ and $Bi_4Si_3O_{12}$*, J. Appl. Phys. **43** (1972), 5110.

4. R. B. Chesler and J. E. Geusic, *Laser Handbook*, vol 1, North Holland (1972), p 362.

5. Z. H. Cho, O. Nalcioglu, and M. R. Farukhi, *Analysis of a Cylindrical Hybrid Position Camera with Bismuth Germanate (BGO) Scintillation Crystals*, IEEE Trans. Nucl. Sci. **NS-25** (1978), 952.

6. M. Couzi, J. R. Vignalou, and G. Boulon, *Infrared and Raman Study of the Optical Phonons in $Bi_4Ge_3O_{12}$ Single Crystals*, Solid State Commun. **20** (1976), 461.

7. S. K. Dickinson, R. M. Hilton, and H. G. Lipson, *Czochralski Synthesis and Properties of Rare-Earth-Doped Bismuth Germanate ($Bi_4Ge_3O_{12}$)*, Mater. Res. Bull. **7** (1972), 181.

8. P. Fischer and F. Waldner, *Comparison of Neutron Diffraction and EPR Results on the Cubic Crystal Structures of Piezoelectric $Bi_4Y_3O_{12}$ (Y = Ge, Si)*, Solid State Commun. **44** (1982), 657.

9.   V. A. Gusev, S. I. Demenko, and S. A. Petrov, *Photoconductivity of $Bi_4Ge_3O_{12}$ Single Crystals*, Optoelectron. Instrum. Data Process. **5** (1988), 32.

10.   V. A. Gusev and S. A. Petrov, *Photoluminescence of $Bi_4Ge_3O_{12}$ Single Crystals*, Optoelectron. Instrum. Data Process. **5** (1988), 15.

11.   S. Haussühl and W. Effgen, *Faraday Effect in Cubic Crystals*, Z. Kristallogr. **183** (1988), 153.

12.   V. Y. Ivanov, A. V. Kruzhalov, V. A. Pustovarov, and V. L. Petrov, *Electron Excitation and Luminescence in $Bi_4Ge_3O_{12}$ and $Bi_4Si_3O_{12}$ Crystals*, Nucl. Inst. Methods Phys. Res. **A261** (1987), 150.

13.   L. F. Johnson and A. A. Ballman, *Coherent Emission from Rare Earth Ions in Electro-Optic Crystals*, J. Appl. Phys. **40** (1969), 297.

14.   A. A. Kaminskii, S. E. Sarkisov, T. I. Butaeva, G. A. Denisenko, B. Hermoneit, J. Bohm, W. Grosskreutz, and D. Schultz, *Growth, Spectroscopy, and Stimulated Emission of Cubic $Bi_4Ge_3O_{12}$ Crystals Doped with Dy^{3+}, Ho^{3+}, Er^{3+}, Tm^{3+}, or Yb^{3+} Ions*, Phys. Status Solidi **(a)56** (1979), 725.

15.   A. A. Kaminskii, D. Schultz, B. Hermoneit, S. Sarkisov, L. Li, J. Bohm, P. Reiche, R. Ehlert, A. A. Mayer, V. A. Lomonov, and V. A. Balashov, *Spectroscopic Properties and Stimulated Emission in the $^4F_{3/2} \to {}^4I_{11/2}$ and $^4F_{3/2} \to {}^4I_{13/2}$ Transitions of Nd^{3+} Ions from Cubic $Bi_4Ge_3O_{12}$ Crystals*, Phys. Status Solidi **(a) 33** (1976), 737.

16.   J. Liebertz, *Einkristallzüchtung von Wismutgermanat ($Bi_4(GeO_4)_3$)*, J. Cryst. Growth **5** (1969), 150.

17.   F. J. Lopez, E. Moya, and C. Zaldo, *Characterization of Chromium Impurities in $Bi_4Ge_3O_{12}$ Single Crystals*, Solid State Commun. **76** (1990), 1169.

18.   C. A. Morrison, *Possible Hosts for Quadruply Ionized $3d^N$ Transition Metal Ions: Na_2TiSiO_5, Y_2SiBeO_7, $Bi_4X_3O_{12}$, and $Bi_{12}XO_{20}$ (X = Si, Ge)*, Harry Diamond Laboratories, HDL-TM-91-1 (April 1991).

19.   C. A. Morrison and R. P. Leavitt, *Crystal Field Analysis of Nd^{3+} and Er^{3+} in $Bi_4Ge_3O_{12}$*, J. Chem. Phys. **74** (1981), 25.

20.   D. P. Neikirk and R. C. Powell, *Laser Time-Resolved Spectroscopy Studies of Host-Sensitized Energy Transfer in $Bi_4Ge_3O_{12}$: Er^{3+} Crystals*, J. Lumin. **20** (1976), 261.

21.   R. Nitsche, *Crystal Growth and Electro-Optic Effect of Bismuth Germanate, $Bi_4(GeO_4)_3$*, J. Appl. Phys. **36** (1965), 2358.

22. G. P. Pazzi, P. Fabeni and R. Linari, *Bismuth Germanate Growth and Properties*, 14th Congress of the International Commission for Optics, H. H. Arsenault, ed., Proc. SPIE **813** (24–28 August 1987), 247.

23. H. Von Philipsborn, *Croissance d'eulytine $Bi_4Si_3O_{12}$ et des composés substitués $Bi_4Ge_3O_{12}$ par la méthode Czochrolski*, J. Cryst. Growth **11** (1971), 348.

24. S. E. Sarkisov, K. K. Pukhov, A. A. Kaminskii, A. G. Petrosyan, and T. I. Betaeva, *Manifestation of Electron-Phonon Interaction in Insulating Crystals Doped with Pr^{3+} Ions*, Phys. Status Solidi **(a)113** (1989), 193.

25. H. Schweppe, *Electromechanical Properties of Bismuth Germanate $Bi_4(GeO_4)_3$*, IEEE Trans. Sonics Ultrasonics (October 1969), 219.

26. D. J. Segal, R. P. Santoro, and R. E. Newnham, *Neutron-Diffraction Study of $Bi_4Si_3O_{12}$*, Z. Krist. **123** (1966), 73.

27. F. Smet, P. Bennema, J. P. Van DerEerden, and W.J.P. Van Enckevort, *Crystal Morphology of Bismuth Germanate ($Bi_4Ge_3O_{12}$)*, J. Cryst. Growth **97** (1989), 430.

28. M. J. Weber and R. R. Monchamp, *Luminescence of $Bi_4Ge_3O_{12}$: Spectral and Decay Properties*, J. Appl. Phys. **44** (1973), 5495.

29. R.W.G. Wyckoff, *Crystal Structures*, vol 4, Interscience, New York (1968), p 172.

49. $Bi_{12}XO_{20}$ (X = Ge, Si, Ti)

49.1 Crystallographic Data on $Bi_{12}XO_{20}$

Cubic T^3 ($I23$), 197, $Z = 2$

Ion	Site	Symmetry	x	y	z	q	α (Å^3)
Bi	24f	C_1	x	y	z	3	2.23[a]
X	2a	T	0	0	0	4	$\alpha_x{}^a$
O_1	24f	C_1	x	y	z	-2	1.349[b]
O_2	8c	C_3	x	x	x	-2	1.349
O_3	8c	C_3	x	x	x	-2	1.349

[a]Fraga et al (1976).
[b]Schmidt et al (1979).

49.2 X-Ray Data on $Bi_{12}XO_{20}$

X	a	x_{Bi}	y_{Bi}	z_{Bi}	x_{O1}	y_{O1}	z_{O1}
Ge	10.1455[a]	0.82409	0.68158	0.98433	0.8655	0.7477	0.5145
Si	10.10433[b]	0.17564	0.31741	0.01592	0.1348	0.2523	0.4858
Ti	10.188[c]	0.17648	0.31862	0.01546	0.1341	0.2511	0.4881

X	x_{O2}	x_{O3}	$\alpha_x{}^d$
Ge	0.8019	0.0977	0.12
Si	0.1950	0.9059	0.03
Ti	0.1930	-0.1045	0.505

[a]Abrahams et al (1967).
[b]Abrahams et al (1979).
[c]Swindells and Gonzalez (1988).
[d]Fraga et al (1976).

49.3 Crystal-Field Components, A_{nm} (cm^{-1}/Å^n), for Ge (T) Site of $Bi_{12}GeO_{20}{}^a$

A_{nm}	Monopole	Self-induced	Dipole	Total
A_{32}	$i54{,}550$	$-i17{,}875$	$i50{,}303$	$i86{,}978$
A_{40}	$-25{,}350$	$13{,}407$	$-26{,}296$	$-38{,}240$
A_{60}	$4{,}766.4$	$-5{,}328.1$	$7{,}328.2$	$6{,}766.5$
A_{62}	163.03	17.182	-35.978	144.24
A_{72}	$-i3901.1$	$i5796.2$	$-i6963.0$	$-i5067.9$

[a]$A_{44} = \sqrt{5/14}\ A_{40}$, $A_{64} = -\sqrt{7/2}\ A_{60}$, $A_{66} = -\sqrt{5/11}\ A_{62}$, and $A_{76} = \sqrt{11/13}\ A_{72}$.

49.4 Crystal-Field Components, A_{nm} (cm^{-1}/Å^n), for Si (T) Site of $Bi_{12}SiO_{20}{}^a$

A_{nm}	Monopole	Self-induced	Dipole	Total
A_{32}	$-i64,514$	$i23,949$	$-i64,411$	$-i104,976$
A_{40}	$-31,029$	$18,698$	$-35,224$	$-47,555$
A_{60}	$6,361.3$	$-8,079.2$	$10,613$	$8,895.5$
A_{62}	170.64	18.406	-35.384	153.67
A_{72}	$i5,446.7$	$-i9,160.9$	$i10,524$	$i6,810.0$

^{a}See footnote a in table 49.3.

49.5 Crystal-Field Components, A_{nm} (cm^{-1}/Å^n), for Ti (T) Site of $Bi_{12}GeO_{20}{}^a$

A_{nm}	Monopole	Self-induced	Dipole	Total
A_{32}	$-i40,105$	$i10,756$	$-i31,746$	$-i61,095$
A_{40}	$-18,128$	$7,592$	$-15,178$	$-25,713$
A_{60}	$2,922$	$-2,608$	$3,736$	$4,050$
A_{62}	156.7	16.29	-32.01	141.0
A_{72}	$i2,192$	$-i2,639$	$i3,292$	$12,844$

^{a}See footnote a in table 49.3.

49.6 Bibliography and References

1. S. C. Abrahams, J. L. Bernstein, and C. Svensson, *Crystal Structure and Absolute Piezoelectric d_{14} Coefficient in Laevorotatory $Bi_{12}SiO_{20}$*, J. Chem. Phys. **71** (1979), 788.

2. S. C. Abrahams, P. B. Jamieson, and J. L. Bernstein, *Crystal Structure of Piezoelectric Bismuth Germanium Oxide $Bi_{12}GeO_{20}$*, J. Chem. Phys. **47** (1967), 4034.

3. G. Barbonas, D. Senuliene, A. Sileika, E. I. Leonov, and V. M. Orlov, *Induced Circular Dichroism in Bismuth Silicate ($Bi_{12}SiO_{20}$) Crystals Doped with Cr*, Liet. Fiz. Rinkinys. **27** (1987), 318 (in Russian, English abstract).

4. S. Denagbe, M. Martin, M. Schvocrer, F. Marsaud, J. C. Launay, and P. Hagenmuller, *Thermally Stimulated Luminescence in Undoped and Doped $Bi_{12}GeO_{20}$ Single Crystals*, J. Phys. Chem. Solids **51** (1990), 171.

5. S. Fraga, K. M. Saxena, and J. Karwowski, *Physical Science Data: 5. Handbook of Atomic Data*, Elsevier, New York (1976).

6. N. I. Katsavets, L. B. Kuleva, E. I. Leonov, I. P. Nikitina, and O. V. Titkova, *Optoelectronic and Structural Properties of the Surface of $Bi_{12}GeO_{20}$ and $Bi_{12}SiO_{20}$ Single Crystals*, Sov. Phys. Tech. Phys. **34** (1989), 1435.

7.   Li Zengfa, Su Dazhao, Yu Xiaoyan, Song Sufa, and Zhang Wanlin, *Study of the Optical Properties of $Bi_{12}SiO_{20}$*, Infrared Phys. **31** (1991), 59.

8.   C. A. Morrison, *Possible Hosts for Quadruply Ionized $3d^N$ Transition Metal Ions: Na_2TiSiO_5, $Y_2SiBe_2O_7$, $Bi_4X_3O_{12}$, and $Bi_{12}XO_{20}$ (X = Si, Ge)*, Harry Diamond Laboratories, HDL-TM-91-1 (April 1991).

9.   C. A. Morrison, *Host Materials for $4d^N$ and $5d^N$ Transition-Metal Ions*, Harry Diamond Laboratories, HDL-TM-90-20 (December 1990).

10.   E. G. Nikolova and M. M. Gospodinov, *The Twyman Effect in Crystals of $Bi_4Ge_3O_{12}$ and $Bi_{12}SiO_{20}$*, J. Mater. Sci. Lett. **8** (1989), 309.

11.   P. C. Schmidt, A. Weiss, and T. P. Das, *Effect of Crystal Fields and Self-Consistency on Dipole and Quadrupole Polarizabilities of Closed-Shell Ions*, Phys. Rev. **B19** (1979), 5525.

12.   D. Senulience, G. Babonas, A. Sileika, E. Leonovas, and V. Orlovas, *An Investigation of the Absorption and Circular Dichroism Spectra in $Bi_{12}SiO_{20}$ Crystals Doped with Ni, Co, and Mn*, Sov. Phys. Collection **27** (1987), 55.

13.   D.C.N. Swindells and J. L. Gonzalez, *Absolute Configuration and Optical Activity of Laevoratatory $Bi_{12}TiO_{20}$*, Acta Crystallogr. **B44** (1988), 12.

14.   W. Wardzynski, M. Baran, and H. Szymczak, *Electron Paramagnetic Resonance of Fe^{3+} in Bismuth Germanium Oxide Single Crystals*, Physica **111B** (1981), 47.

15.   W. Wadzynski, T. Lukasiewicz, and J. Zmija, *Reversible Photochromic Effects in Doped Single Crystals of Bismuth Germinate ($Bi_{12}GeO_{20}$) and Bismuth Silicon Oxide ($Bi_{12}SiO_{20}$)*, Optics Commun. **30** (1979), 203.

16   W. Wardzynski and H. Szymczak, *The Center of Orthorhombic Symmetry in Chromium Doped $Bi_{12}GeO_{20}$ and $Bi_{12}SiO_{20}$ Single Crystals*, J. Phys. Chem. Solids. **45** (1984), 887.

17.   W. Wardzynski, H. Szymczak, K. Pataj, T. Lukasiewicz, and J. Zmija, *Light Induced Charge Transfer in Cr Doped $Bi_{12}GeO_{20}$ and $Bi_{12}SiO_{20}$ Single Crystals*, J. Phys. Chem. Solids **43** (1982), 767.

18.   M. J. Weber, ed., *CRC Handbook of Laser Science and Technology*, vol V part 3, vol IV part 2, CRC Press, Boca Raton, FL (1987).

19.   W. Wojdowski, T. Lukasiewicz, W. Nazarewicz, and J. Zmija, *Infrared Studies of Lattice Vibrations in $Bi_{12}GeO_{20}$ and $Bi_{12}SiO_{20}$ Crystals*, Phys. Status Solidi **(b)94** (1979), 649.

Acknowledgements

I owe a special thanks to D. S. McClure who during the early phases of this work provided me with results and advice which had a lasting influence on this work. The following individuals provided me with unpublished data, results before publication, and useful information: J. B. Gruber, M. E. Hills, P. Porcher, T. H. Allik, J. A. Capobianco, B. Aull, D. Singel, H. P. Jenssen, L. Merkle, H. Verdun, F. S. Richardson, E. D. Filer, N. P. Barnes, and F. S. Bartram. This work was influenced by discussions with L. Esterowitz, A. Pinto, R. Powell, G. Quarles, T. Allik, C. K. Jorgensen, and R. Reisfold. C. K. Jorgensen also supplied me with many references and reprints to earlier work on the $4d^N$ and $5d^N$ transition-metal ions.

This work would have been impossible without the capable technical assistance of my coworkers, D. E. Wortman, R. P. Leavitt, and J. D. Bruno; their discussions and their computer skills were invaluable. Other coworkers from whom I have received considerable information and technical assistance are S. B. Stevens, M. S. Tobin, T. B. Bahder, and G. A. Turner. I would like to thank the high school summer students, Betsy, Ken, and Suzie Wong, Ron, Glen, and Wayne Lee, Mandy Hansen, and Barry Reich; their cheerful attitude and readiness to learn made working with them a pleasure.

I have received much support (both financial and moral) from Harry Diamond Laboratories (J. Reed and J. Sattler), the Naval Research Laboratory (L. Esterowitz), the Center for Night Vision and Electrophysics (R. Buser and A. Pinto), and the NASA Langley Research Center (N. Barnes and E. Filer). I would like to recognize the HDL library personnel, B. McLaughlin, P. Dore, L. Merson, and the late N. Brandt for their cheerful assistance in acquiring the many references. Finally, I thank Ann Fultz for her excellent typing of the manuscript and Barbara Collier for the outstanding work on all phases of the manuscript, page arrangements, and layout of the tables and overcoming the almost insurmountable barrier of transferring information between different word processors.

Index

Italicized page numbers refer to literature citations.

A

Ag. *See* silver
A_{nm}. *See* crystal-field components, A_{nm}
Au. *See* gold

B

Ballhausen parameters
  Ds, Dt, and Dq 11
  $D\sigma$, $D\tau$, and Dq 12
B_{nm}. *See* crystal-field parameters, B_{nm}

C

cadmium
  Cd^{3+}
    free ion 19
    Hartree-Fock 25
  $Cd4+$ 157
    free ion 19
    Hartree-Fock 25
Cd. *See* cadmium
chromium
  Cr^{2+} 51, 52, *53*, *54*, 55, 88, *89*, 92,
    109, 150, *151*, *154*, *155*
    free ion 17
    Hartree-Fock 24
  Cr^{3+} *14*, 27, 28, *29*, *30*, *31*, 33, *34*,
    35, *39*, *40*, 42, *43*, 52, *54*, 56, 57,
    58, *59*, 62, 63, *64*, 69, 70, *73*, *74*,
    75, *76*, 77, 78, 81, *82*, 97, 98, 104,
    106, *108*, *109*, 110, 111, *112*, *113*,
    116, *117*, *118*, *119*, *120*, 130, 134,
    150
    free ion 17
    Hartree-Fock 24
  Cr^{4+} 37, *154*
    free ion 17
    Hartree-Fock 24

cobalt
  Co^{2+} 29, *31*, 42, 46, *49*, 51, 52, *55*,
    56, 68, *69*, *70*, *71*, 77, 78, 84, *85*,
    86, 88, *90*, 92, 97, 98, 111, *112*,
    113, 130, 134, *150*, *153*
    free ion 17
    Hartree-Fock 24
  Co^{3+} 27, 28, 29, *31*, 33, 57, 62, 63,
    98, 104, *107*, 130, 134, 150
    free ion 17
    Hartree-Fock 24
  Co^{4+} 37
    free ion 17
    Hartree-Fock 24
copper
  Cu^{2+} 51, *53*, 58, *86*, 92, *113*
    free ion 18
    Hartree-Fock 24
  Cu^{3+} 27, 28, 33, 57, 62, 63, 84, 99,
    104, *105*, 150
    free ion 18
    Hartree-Fock 24
  Cu^{4+} 37
    free ion 17
    Hartree-Fock 24
Cr. *See* chromium
crystal-field components, A_{nm} 6, 8, 9,
    10, 12, 13, 23
  monopole 8
  nonvanishing B_{nm} and A_{nm} 6, 7, 23
  dipole and self-induced 8, 9
crystal-field parameters, B_{nm} 2, 6, 9,
    10, 11, 12, 13, 23
Cu. *See* copper
cubic sites
  octahedral 8